Barthélemy BAWAR

Monitoring periodic pollution of water resources in Burkina Faso

Barthélemy BAWAR

Monitoring periodic pollution of water resources in Burkina Faso

Monitoring by the DGRE laboratory between April 2023 and April 2024

ScienciaScripts

Cover image: www.ingimage.com

This book is a translation from the original published under ISBN 978-620-6-71197-1.

Publisher:
Sciencia Scripts
is a trademark of
Dodo Books Indian Ocean Ltd. and OmniScriptum S.R.L publishing group

120 High Road, East Finchley, London, N2 9ED, United Kingdom
Str. Armeneasca 28/1, office 1, Chisinau MD-2012, Republic of Moldova, Europe
Managing Directors: Ieva Konstantinova, Victoria Ursu
info@omniscriptum.com

Printed at: see last page
ISBN: 978-620-8-52803-4

Contents

PART I

BURKINA FASO

Unité - Progrès - Justice

MINISTERE DE L'ENVIRONNEMENT, DE L'EAU ET DE L'ASSAINISSEMENT

SECRETARIAT GENERAL

GENERAL SECRETARIAT

Unity - Progress - Justice

MINISTRY OF THE ENVIRONMENT, WATER AND SANITATION

MISSION SUMMARY REPORT

INVESTIGATION INTO THE DEATH OF
OF FISH IN THE YAOGHIN DAM
YAOGHIN DAM, POA COMMUNE, BOULKIEMDE PROVINCE
BOULKIEMDE, CENTRE-WEST REGION

April 2023

INTRODUCTION

In Burkina Faso, the mobilisation and development of water resources is one of the major thrusts of the sustainable development policy. To this end, a number of projects and programmes have been and continue to be implemented to harness groundwater and surface water resources. However, faced with the dual challenge of increasing production and combating climatic conditions, beneficiaries are integrating the use of inputs into agriculture. Although necessary to improve production, the misuse of organic or mineral fertilisers and pesticides is proving dangerous for the aquatic ecosystem. Carried along by run-off water, mineral surpluses accumulate in surface water reservoirs, leading to eutrophication. In this process, phosphate, nitrogen and sulphate inputs are the most significant. There are also other risks, such as water pollution and pathogen contamination caused by flooding and higher concentrations of pollutants during droughts.

The pollution of water resources has worsened and is becoming increasingly recurrent in Burkina Faso with the use of chemicals such as cyanides at mining and gold-panning sites. Numerous incidents of pollution have been reported, including

- the pollution incident at the Youga mine site in August 2008;
- the pollution incident at the Nobsin gold panning site in July 2009 ;
- the pollution incident in the Nakanbé: the case of the Ziga dam in 2012 ;
- the pollution incident at the Guiti dam in May 2020;
- the pollution incident at the gold-panning site in the commune of Siby July 2021;
- the death around 200 Dafra catfish in the Houet marigot on 24 February 2023,
- the pollution incident around the Mouhoun River between Secaco and Boromo in February 2023.

These pollution incidents have led to death of animals and fish. This recurrence of pollution is a call to improve the monitoring of both the quality and quantity of water resources in order to better protect them.

Recently, the governor of the Centre-West region sent a memo to the people of Poa concerning cases of dead fish at the Yaoghin dam since the morning of Thursday 30 March 2023. Market gardening activities are carried out around the dam by local people, who use pesticides and fertilisers that are likely to pollute the water in the dam.

On the instructions of the Secretary General of the Ministry of the Environment, Water and Sanitation, a series of joint missions led by the Directorate General for Water Resources (DGRE) and the Directorate General for Environmental Preservation (DGPE) visited the Centre-West Region, more specifically Yaoghin in the commune of Poa, on 31 March 2023 and again on 19 April 2023, to carry out technical investigations into the fish mortality that had occurred in the dam in the said village.

The specific objectives of the missions were to :

For the first mission :

- Meet the authorities at regional and/or local level and inform them of the objective of the mission;
- Take stock of the pollution of the water in the dam;
- Define lines of enquiry ;
- Carry out on-site analyses and take water samples for in-depth laboratory analysis;
- Speculate on the nature of the pollution in the water from the dam.

For the second mission :

- changes in water quality at the dam at least once;
- Check the hypotheses about the nature of the pollution of the water in the dam;
- Make recommendations to limit any pollution of water resources.

The expected results were :

For the first mission :

- The authorities at regional and/or local level are met and informed of the purpose of the mission;
- Findings on the state of the dam's water pollution were made;
- In-situ analyses and taking water samples for analysis

are carried out in the laboratory;

- Hypotheses were put forward as to the nature of the pollution of the water in the dam.

For the second mission :

- A mission to monitor changes in the quality of the dam is being carried out;
- The hypotheses concerning the nature of the pollution of the water in the dam have been verified;
- Recommendations have been made to limit any pollution of water resources.

1. OVERVIEW OF JOINT MISSIONS

1.1. Course of the first joint mission

The team **first** travelled to Koudougou, where it met the acting Regional Director of Water and Sanitation, Mr OUEDRAOGO Daouda, to discuss the purpose of the mission and hear his version of events.

Our discussions enabled us to gather the following information:

- A joint mission from the Direction Régionale de l'Eau et de l'Assainissement du Centre-Ouest (DREA-CO), the Agence de l'Eau du Mouhoun and the Chef de Zone d'Appui Technique de l'Elevage (ZATE) visited the Yaoghin dam on 30 March 2023 to observe the fish mortality and take water samples for possible analysis;
- a large quantity of water entering the dam reservoir as a result of heavy rainfall on 28 March 2023, leading to an increase in the turbidity of the water in the dam;
- The first signs of fish dying and dying were observed on the evening of 29 March 2023 ;
- The number of dead fish increased on the morning of 30 March 2023. To prevent human poisoning, ZATE officials seized the dead and dying fish;
- samples of dead fish were taken by the POA veterinary station for analysis;
- the Direction Générale des Ressources Halieutiques (DGRH) instructed the ZATE managers to take a sample of mud from the bowl and a dying fish preserved in alcohol (to prevent the dead fish from rotting);
- market gardening around the dam;
- the water level in the dam was very low before the incident;
- the absence of gold panning in the locality;
- the incident is a first for the dam
- two water samples were taken, one on the left bank and the other on the right bank by DREA-CO agents, which we recovered for in-depth laboratory analysis.

Secondly, the mission went to the Poa town hall to inform them of the purpose of the mission. At this point, the mission met with a mission from the National Public Health Laboratory, which had come for the same purpose.

Following the interviews, the mission travelled to the dam site, accompanied by Mr Justin KABRE, a Poa Town Hall official, to carry out observations and take water samples for

analysis.
Once on site, we found the Head of the Livestock Technical Support Zone (ZATE). Discussions with the Head of ZATE enabled us to gather the same information as that obtained from the Regional Water and Sanitation Department.
Observations and hearings in the field revealed the following:
- The dead fish in the reservoir are mainly sardines and carp of the Cyprinidae family, with very few catfish, according to local residents;
- the water looks greyish;
- the presence of animal waste in the basin of the dam, particularly on the banks;
- the water from the dam gives off a nauseating odour;
- a large quantity of plant debris carried by rainwater and found on the banks;
- The dead fish ended up on the left bank, probably because of the direction of the wind, which blew from the right bank towards the left;
- an absence of invasive plants in the dam lake;
- the absence of gold-panning sites in the vicinity of the reservoir;
- a high density of market gardening on the banks of the dam and all along the watercourse, despite the measures occupation;
- Jatropha cultivation all around the dam and in large quantities on the left bank of the dam;
- clay brick production areas in the dam basin.

With these questions in mind, some local residents are accusing the market gardeners of using highly toxic products to treat their crops. When it rained in the locality on 28 March 2023, the water carried the pesticide residues into the dam via the canal linking the gardens to the dam.
After these routine observations, the team took water samples and carried out *in situ* measurements on both banks and on the dam's embankment. Four sampling points (Appendix 3) were identified and coded (P1: left bank, P2: dike1, P3: dike2 and P4: right bank). Their GPS coordinates were taken.
The water samples were taken in accordance with the Burkinabe standards "NBF 08-003-1 to 4: 2018" relating to water sampling, including surface water.
The water samples taken are kept cool in coolers. Some are then sent to the DGRE's Raw Water Analysis Laboratory (to analyse additional parameters such as major ions and trace metals), while others are sent to the Environmental Quality Analysis Laboratory (for BOD5 and COD analyses). In addition, it would be useful to have the results of analyses carried out by other players who were also involved, in particular the ministry responsible for fisheries resources through the DGRH and the ministry responsible for health through the Agence Nationale de Sécurité Sanitaire de l'Environnement, de l'Alimentation et du Travail et des produits de santé (ANSSEAT, formerly LNSP) for more comprehensive technical investigations.

1.2. Course of the second mission

The mission team first went to the Poa commune town hall to inform them of the purpose of the mission.
After a meeting with the Secretary General of Poa Town Hall, the mission travelled to the dam site, accompanied by Mr Justin KABRE, an agent of Poa Town Hall, to make observations and take water samples for analysis.
Observations and hearings in the field revealed the following:
- No more dead fish at the dam

- The water in the dam is more or less clear and limpid;
- the presence of animal waste in the basin of the dam, particularly on the banks;
- the water from the dam no longer emits a foul odour;
- the use of water by market gardeners through the installation of swivel pumps on the left bank and dyke,
- the presence of a health (CSPS Yaoghin) 200 metres from the dam...

After these routine observations, the team took water samples and carried out *in situ* measurements on both banks and on the dam embankment, at the same sampling points as the first mission.

The water samples were taken in accordance with the Burkinabe standard "NBF 08-003-1 to 4: 2018" relating to surface water sampling.

Water samples were kept in coolers and sent to the DGRE's Laboratoire d'Analyses des Eaux brutes (LAEB) and the Laboratoire d'Analyses de la Qualité de l'Environnement (LAQE) for further analysis.

Another part of the samples (04 samples) was to be sent to the National Public Health Laboratory (LNSP) for pesticide residue analysis.

2. RESULTS AND INTERPRETATION

2.1. In situ analysis

The in situ measurements covered the following parameters: temperature, pH, conductivity, turbidity and dissolved oxygen.

2.2. Laboratory analysis

Laboratory analyses were carried out on :

- Major ions: chlorides (Cl^-), nitrates (NO_3^-), nitrites (NO_2-), orthophosphates (HPO_4^{2}-), sulphates (SO_4^{2}-) as well as sodium (Na) and potassium (K).
- Cyanides and/or pesticides.
- Metalloids (metallic trace elements): copper (Cu), zinc (Zn), manganese (Mn).
- Chemical indices (Biological Oxygen Demand (BOD_5) and Chemical Oxygen Demand (COD)).

2.3. Analysis equipment and methods used

The equipment and methods used to analyse the parameters studied are listed in the table in Appendix 6.

2.4. Presentation of analysis results

Table 1 gives the results of in-tu and laboratory analyses of several physico-chemical parameters for the first mission.

Table 2 gives the results of in-tu and laboratory analyses of several environmental parameters for the second mission.

Table I: Results of in situ measurements and laboratory analyses for the first mission

Samples		In situ measurements					Laboratory analysis												
Sample code	**Geographical location**	T(°C)	pH	Conductivity (ц3/cr)	Turbidity (NTU)	Dissolved oxygen (mg/l)	cr (mg/l)	NO_3- (mg/l)	NO_2- (mg/l)	PO?' (mg/l)	SO4 [12] - (mg/l)	Na (mg/l)	K (mg/l)	Cu (mg/l)	Mn (mg/l)	Zn (mg/l)	GN' (mg/l)	COD (mg/l)	DB 05 (mg/l)
PI (Rive Gauche)	12°13'32,2 "N 002°06'15,7 "W	30,3	6,51	113,0	722,6	0,54	2,6	0,5	0,31	0,1	0,6	2,1	16,3	<0,05	3,09	0,11	0,038	146	136
P2 (Dyke)	12°13'29,0 "N 002°06'12,5 "W	28,0	6,3	111,0	712,2	0,81	2,4	<0,1	0,22	0,1	0,7	2,2	18,2	<0,05	2,89	0,05	0,015	116	104

[1] Decree n°2001-185/PRES/PM/MEE setting the limits for pollutant discharges into fair, water and soil, defining the limits for the protection of fish water quality

[2] Joint decree of the Ministry of Planning, Water and Environment n°2028-03 (5 November 2003) setting the quality standards for fish waters in Morocco.

P3 (Dyke)	12°13'25,5 "N 002°06'12,0 "W	27,9	6,22	110,4	718,0	0,48	2,4	0,1	0,33	0,08	0,8	2,1	15,2	<0,05	2,89	0,10	0,016	125	99
P4 (Right Bank)	12°13'21,7 "N 002°06'll,5 "W	28,0	6,32	109,3	705,5	0,44	2,4	<0,1	0,6	0,1	0,7	2,1	13,8	<0,05	2,89	0,09	0,027	138	107
Average values		28,5	6,33	110,9	714,6	0,57	2,45	0,2	0,37	0,08	0,7	2,125	15,875	0,05	2,94	0,09	0,024	131	112
National standards[1]		40°C	6 à 10,3	1000	-	>7								0,3		<1			<6
Moroccan standards[2]		8 à 30	5 à9	<3000	-	>3	-	-	<0,5	-	<200	-	-	<40	<0,1	< 1,3	<0,05	<30	<6

Table II: Results of in situ measurements and laboratory analyses of water samples from the second mission

Samples		In situ measurements					Laboratory analysis											
Sample code	Geographical location	T (°C)	pH	Conductivity (pS/cm)	Turbidity (NTU)	Dissolved oxygen (mg/l)	Ci (mg/l)	NO3 (mg/l)	PO4[3] (mg/l)	SO4[2] (mg/l)	Na (mg/l)	K (mg/l)	Cu (mg/l)	Mn (mg/l)	Zn (mg/l)	COD (mg/l)	BOD 5 (mg/l)
PI (Rive Gauche)	12°13'32,2 "N 002°06'15,7 "W	32,8	6,50	137,9	56,54	8,66	3,6	<0,1	<0,1	0,6	<0,05	8,34	0,054	0,052	0,104	8	6
P2 (Dyke)	12°13'29,0 "N 002°06'12,5 "W	32,4	6,73	139,2	44,22	8,62	3,4	<0,1	<0,1	0,6	<0,05	2,62	<0,05	0,069	0,067	6	5
P3 (Dyke)	12°13'25,5 "N 002°06'12.0 "W	31,4	6,88	139,4	57,98	8,42	3,4	<0,1	<0,1	0,5	<0,05	<0,05	<0,05	0,072	<0,05	12	9
Samples		In situ measurements					Laboratory analysis											
Code samples	Location	T (°C)	pH	Conductivity (pS/cm)	Turbidity (NTU)	Dissolved oxygenne (mg/1)	Cl (mg/1)	NO3 (mg/1)	PO4[3] (mg/1)	SO4[2] (mg/1)	Na (mg/1)	κ (mg/1)	Cu (mg/1)	Mn (mg/1)	Zn (mg/1)	COD (mg/1)	BOD5 (mg/1)
P4 (Right	12°13'21.7 "N	31,9	7,08	138,5	59,84	8,27	3,6	<0,1	<0,1	0,5	<0,05	<0,05	<0,05	0,070	<0,05	10	8
Average values		32,1	6,8	113,75	54,64	8,49	3,5	0,1	0,1	0,6	<0,0	2,76	0,05	0,06	0,04	9	
National standards[3]		40° C	6 à 10,	1000	-	>7							0,3		<1		<6
Moroccan standards[4]		8á 30	5 à 9	<3000	-	>3	-	-	-	200	-	-	<40	<0,1	<1,3	<30	<6

2.5. Interpretation of analysis results

2.5.1. Interpretation of in-situ analysis results

Figure 1 the interpretation of the in situ measurements for the two separate missions. These parameters are temperature (T), pH, conductivity, turbidity and dissolved oxygen in the water.

2.5.1.1. Temperature

The temperature of the water affects its density and viscosity, the solubility of oxygen, the speed of chemical and biochemical reactions, and the rhythms of activity of organisms, most

[3] Decree n°2001-185/PRES/PM/MEE setting the limits for pollutant discharges into fair, water and soil, defining the limits for the protection of fish water quality

[4] Joint decree of the Ministry of Planning, Water and the Environment n°2028-03 (5 November 2003) setting the quality standards for fish waters in Morocco.

of which are poïkilothermic.
Looking at the results, we can see that the average temperature during the first mission was 28.5°C, while during the second mission the average temperature was higher, at 32.12°C. This increase in temperature may be due to various factors, such as climate change, human activity or the sampling period.

2.5.1.2. pH

The pH is the cologarithm of the concentration of hydrogen ions in water: 0 to 7 is acidic and 7 to 14 is alkaline. The pH cannot be dissociated from temperature, dissolved oxygen and mineralisation. Fish generally tolerate a pH of between 6 and 9.
In terms of pH, the average was 6.33 on the first mission and 6.80 on the second. These results are more or less within the acceptable range of national and Moroccan standards, which recommend a pH of between 6 and 10.3 and between 5 and 9, respectively.

2.5.1.3. Conductivity

Electrical conductivity (EC) is a measure of the ability of an aqueous solution to conduct an electric current, which depends on the concentration of ions in solution. It is sensitive to variations in dissolved matter, mainly mineral salts, and can be useful for characterising a body of water. Conductivity also increases with water temperature.
Conductivity, which measures a substance's ability to conduct electricity, averaged 110.9 µS/cm on the first mission and 138.75 µS/cm the second. National standards set a limit of 1000 µS/cm, indicating that both missions were below this limit.

2.5.1.4. Turbidity

The turbidity of the water depends on the nature of the terrain it crosses, the season, the rainfall, the water flow and the nature of the discharges. It can be assessed by the transparency of the water.
In terms of turbidity, the average during the first mission 714.6 NTU, while during the second mission it was much lower, with an average of 54.64 NTU.
Although national standards do not provide a specific limit for surface water turbidity, low turbidity is generally considered to be an indicator of good water quality.

2.5.1.5. Dissolved oxygen

Along with pH values, dissolved oxygen concentrations are one of the most important water quality parameters for aquatic life.

- It generally accounts for 35% of the volume of total gases dissolved in water.
- Air-water exchanges take place through contact, increased by mixing on lotic facies. Photosynthesis increases the dissolved oxygen content. A level of less than 3 mg/l is abnormal.
- It is lethal for cyprinids when it falls below 4 mg/l and 6 mg/l (75% saturation) for salmonids. The vital value for these fish must be greater than 6 mg/l.
- Levels above the natural oxygen saturation level indicate eutrophication of the environment, resulting in intense photosynthetic activity.

Dissolved oxygen is essential for the vast majority of organisms living in water, as the solubility of O2 varies according to temperature, mineralisation and atmospheric pressure.
Finally, dissolved oxygen levels in the water averaged 0.57 mg/l during the first mission and 8.49 mg/l during the second mission. National standards recommend a minimum dissolved oxygen content of over 7 mg/l, which means that the results of the first mission were below this standard, while those of the second mission were within the national guide value. This low oxygenation could explain the death of the fish in this dam.

According to the results of the in-situ measurements taken during the first mission, the Yaoghin dam was suffering from anoxic conditions characterised by a very low dissolved oxygen content, which could lead to the death of certain aquatic organisms such as fish. The results of the second mission show a significant improvement in water quality, with dissolved oxygen levels in line with standards.

These in-situ measurements provide important information on the water quality of the Yaoghin dam studied. They show that water quality varies with time and that certain parameters, such as turbidity and dissolved oxygen content, can have important implications for the life of aquatic organisms and the environment.

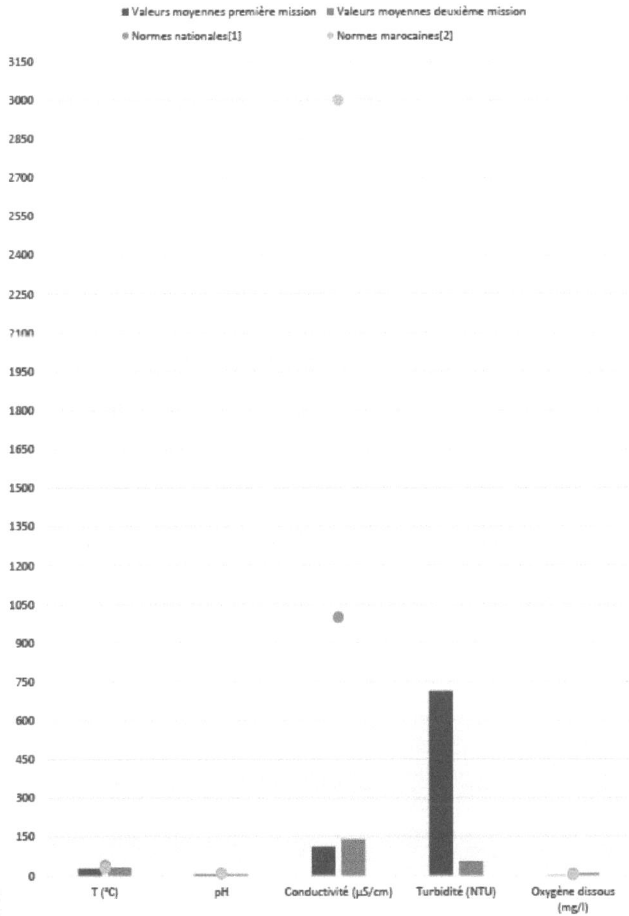

Figure 1: variation over time in the mean values of in-situ parameters

2.5.2 Interpretation of laboratory analysis results

Tables I and II provide the results of analyses for various water samples, for the two separate missions, as well as the reference standards for each parameter measured. The national standards decree does not propose values for certain fish water quality parameters.

The Moroccan normative decree also does not specify limit values for chlorides, nitrates, sodium and potassium).

Here is a possible interpretation of the results:

- **Ions such as chlorides, sulphates, phosphates and nitrates**; there was no significant variation in the mean values of these parameters during the two separate missions;
- **Na (sodium)**: The average values for the first mission are 2.125 mg/L, which is relatively low. For the second mission, the values are below the detection limit (< 0.05 mg/L), indicating a significant absence of sodium in the water samples tested. These values therefore comply with the reference standards for sodium;
- **K (potassium)**: The average values for the first mission are 15.875 mg/L, which is also relatively low. For the second mission, the values are 2.76 mg/L, which is even lower. These values therefore comply with the reference standards for potassium;
- **Cu (copper)**: The average values for the first mission are 0.05 mg/L, which is below the detection limit for the second mission (0.051 mg/L). This indicates that the water samples tested contain little or no copper. Both values comply with the reference standards for copper;
- **Zn (zinc)**: The average values for the two missions are 0.09 mg/L and 0.044 mg/L respectively, which is well below the reference standard for zinc (< 1.3 mg/L). The water samples tested therefore contain little or no zinc;
- **Mn (manganese)**: The mean values for the first mission are 2.94 mg/L, which is relatively high compared with reference standards (< 0.1 mg/L). The average value for the second mission was 0.065 mg/L, which is much lower than for the first mission, and complies with the reference standard;
- **Cyanides and/or pesticides**: The results of the analyses of cyanides (**Table I, for the first mission)** and pesticide residues **(Appendix 4, for the second mission)** do not indicate the presence of these toxic elements in the water samples from the dam studied;
- **COD (chemical oxygen demand)**: The average values for the first mission are 131 mg/L, which is higher than the reference standard (< 30 mg/L). The average values for the second mission were 9 mg/L, which is much lower than for the first mission and complies with the reference standard;
- **BOD5 (5-day Biochemical Oxygen Demand)** : The average values for

The average values for the first mission were 112 mg/L, which is higher than the reference standard (< 6 mg/L). The average values for the second mission are 7 mg/L, which is much lower than for the first mission, but still only slightly higher than the reference standard.

As surface water generally contains less than 0.05mg/l, this information leads to conclude that the heavy rain that fell in the locality on 28 March 2023 must have carried waste in run-off water, polluting the dam with high levels of manganese.

As far as fish mortality is concerned, catfish and sardines are two different species with different needs in terms of water quality.

According to the work of (T. R. Crompton et al., 2004) and (K. G. Linden et al., 2000), water pollution by chemicals such as manganese, as well as high levels of COD and BOD5, can have a negative impact on the health and survival of all fish in an aquatic ecosystem. High concentrations of manganese can also affect the quality and quantity of food available to fish, which can also have an impact on their survival.

Furthermore, according to the work of (K. G. Linden et al., 2000) and (H. S. Suleiman et al., 2006), potassium permanganate ($KMnO_4$) is a powerful oxidant and is often used in water treatment to eliminate organic and inorganic contaminants; as a disinfectant, in particular to

sterilise medical equipment and to treat fungal and bacterial infections. Finally, it is also used as a dye for textiles, leather, paper, ceramics and other materials.
However, high concentrations of potassium permanganate in water can be toxic to fish. Potassium permanganate can reduce the level of dissolved oxygen in the water, which can cause fish to suffocate (H. S. Suleiman et al., 2006). In addition, potassium permanganate can irritate the gills and skin of fish, which can also make them more vulnerable to other environmental stresses and diseases.
It is also important to note that the toxicity of potassium permanganate to fish depends on many factors such as the concentration of the chemical in the water, the duration of exposure of the fish and the sensitivity of different fish species (T. R. Crompton et al., 2004).
Interpreting these analytical data, we can conclude that the water quality measured during the first mission was poor, with low levels dissolved oxygen and high levels of K, Mn, COD and BOD5. In contrast, the levels measured during the second mission are much improved, with higher levels of dissolved oxygen and lower levels of K, Mn, COD and BOD5. This indicates that there is a significant relationship between dissolved oxygen levels and the latter elements, and this is verified by the studies of H. S. Suleiman et al. 2006.
Overall, these results show that the water samples studied may have varying levels of different physicochemical parameters, but that the values are largely within the standards set for each element or compound. However, it is important to regularly monitor the levels of these elements in order to maintain the quality of the water at the dam studied.
National and Moroccan standards can be used as references to assess the quality of fish waters in order to guide actions aimed improving it.

3. RECOMMENDATIONS

The following recommendations (**Table III**) have been formulated in order to limit possible pollution of water resources throughout the country.

Table III: Recommendations for quality management of water resources

N°	Recommendations	Deadline for implementation	Implementation structure
1	Develop and implement a communication plan on the impact of human activities on water quality, the proliferation of invasive plants, and the interaction between the hydrographic network and the pollution of water bodies.	31 December 2023	Water agencies, DREA, DGRE
2	Raise awareness among market gardeners of the need to comply with production standards and, in particular, the use pesticides and synthetic chemical fertilisers	Continue	Water Agencies, Water Police, CLE, Town Halls
3	Create herbaceous strips around the water bodies and then plant non-woody forest plants around these strips to limit pollution of the water bodies.	Continue	Water Agencies
4	Ensure compliance with water regulations by up inspections	Continue	DREA (Water Police)
5	Strengthening water quality control and monitoring systems (water quality monitoring networks, strengthening the technical facilities of DGRE and DGPE laboratories).	Continue	MEEA, DGRE, DGPE
6	Given the recurrence of water pollution events, set up an Interministerial Technical Committee to investigate water pollution under the coordination of the MEEA	31 December 2023	SG/MEEA, DGRE
7	Revise the regulations setting standards for surface water/groundwater in general and fish water in particular	31 December 2025	SG/MEEA, DGRE, ABNORM

	to make them national "NBF" standards to be used for this purpose.		
8	Follow current environmental regulations and expert recommendations for the use of chemicals such as potassium permanganate.	continues	Water police, CLE, town halls

CONCLUSION

The aim of our mission was to determine the nature of any pollution of this water by various pollutants in order to pinpoint the probable causes of the fish mortality. The mission team analysed the various physico-chemical parameters of the dam water *in situ* and in the DGRE and DGPE laboratories.

By interpreting these analytical data, we can conclude that the quality of the water measured during the first mission was poor, with low levels dissolved oxygen and high levels of K, Mn, COD and BOD5. In contrast, the levels measured during the second mission are much improved, with higher levels of dissolved oxygen and lower levels of K, Mn, COD and BOD5. This indicates that there is a significant relationship between dissolved oxygen levels and the latter elements, and this is verified by the studies of H. S. Suleiman et al. 2006.

Overall, these results show that the water samples studied may have varying levels of different physicochemical parameters, but that the values are largely within the standards set for each element or compound. However, it is important to regularly monitor the levels of these elements to maintain the quality of the water at the dam studied.

The water in the dam has now returned to normal and can be used for a variety of purposes without fear.

The pollution observed could be due to the heavy rainfall recorded on 28 March 2023, which washed objects and pollutants into the reservoir.

BIBLIOGRAPHICAL REFERENCES

1. Decree no. 2001-185/PRES/PM/MEE setting standards for pollutant discharges into the air, water and soil, and defining standards for the protection fish water quality;

2. Joint decree of the Ministry of Regional Planning, Water and the Environment n°2028-03 (5 November 2003) setting quality standards for fish waters in Morocco ;

3. T. R. Crompton, Toxicants in Aqueous Ecosystems: A Guide for the Analytical and Environmental Chemist (New York: Wiley, 2004), p. 102 ;

4. K. G. Linden, D. A. Lake, and J. S. Sigmund, "Treatment of Permanganate-Contaminated Groundwater by In Situ Chemical Oxidation," Environmental Science & Technology 34, no. 23 (2000): 4879-4883 ;

5. H. S. Suleiman, N. A. Al-Muhtaseb, and R. A. Al-Zoubi, "Removal of Permanganate from Water by Adsorption on Activated Carbon," Journal of Hazardous Materials 137, no. 1 (2006): 529-537 ;

6. NBF 08-003-1 : 2018 : Water Quality - Sampling - Part 1: Guidelines for the design of sampling programmes and techniques,

7. NBF 08-003-2: 2018: Water quality - Sampling - Part 2: Storage and handling of water samples ;

8. NBF 08-003-4: 2018: Water Quality - Sampling - Part 4: Guidance for sampling the waters of natural and artificial lakes

9. Brouwer, A., (2017). Pesticide Residues in European Freshwater Ecosystems: A Review of Recent Monitoring Data and Regulatory Responses. Environmental Sciences Europe, 29(1), 1-18.

APPENDIX 1: LIST OF AUTHORITIES MET IN THE LOCALITY

Structure	Identity
Acting Regional Director of Water and Sanitation for the Centre-West region	OUEDRAOGO Daouda Tel : 76 60 40 74
Head of the Matter Transferred Department at Poa Town Hall	Justin KABRE Tel : 76 62 43 78
the Head of the Livestock Technical Support Zone (ZATE)	KABORE Issouf Tel : 76 67 39 21

APPENDIX 2: ILLUSTRATIVE PHOTOS, DGRE, MARCH 2023

Photo 1: dead fish Photo 2: dead catfish

Figure 3: Water level at the dam Figure 4: Measurement of in situ parameters

Figure 5: Plant debris carried by rainwater Figure 6: Jatropha plant near the dam

APPENDIX 3: ILLUSTRATIVE PHOTOS, DGRE, APRIL 2023

Figure 1: Aspect of the water during the second mission
Figure 2: Market gardeners drawing water from pivot pumps

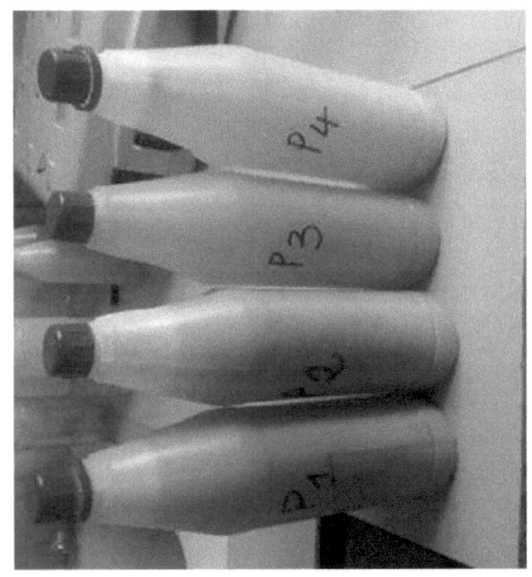

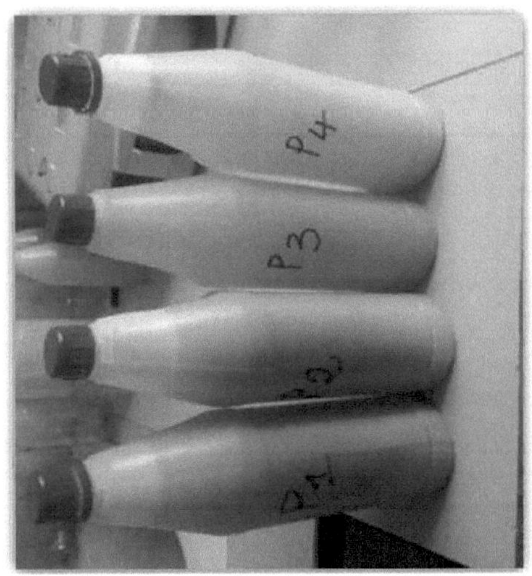

Figure 3: the four water taken-1 Figure 4: the four water samples taken-2

APPENDIX 4: TABLE I: RESULTS OF PESTICIDE ANALYSES OF WATER SAMPLES FROM MISSION 2

Ж)		ORGANOCHLORES															
	Parameters	**2,4DDT**	**Linen**	**Aid**	**Op'DDT**	**Met**	**Mir**	**PCB 209**	**Diel**	**Hep**	**Alpha-endosulfan**	**Beta-endosulfan**	**BCB**	**Chi**	**Cldo**	**meth**	**Pre**
	Units	**mg/L**	**mg/L**	**mg/L**	**mg/L**	**mg/L**	**mg/L**	**mg/L**	**mg/L**	**mg/L**	**mg/L**	**mg/L**	**mg/L**	**mg/L**	**mg/L**	**mg/L**	**mg/L**
	PI (Rive Gauche)	<0,01	<0,01	<0,01	<0,01	<0,01	<0,01	<0,01	<0,01	<0,01	<0,01	<0,01	<0,01	<0,01	<0,01	<0,01	<0,01
Codes samples	*P2 (Dyke)*	<0,01	<0,01	<0,01	<0,01	<0,01	<0,01	<0,01	<0,01	<0,01	<0,01	<0,01	<0,01	<0,01	<0,01	<0,01	<0,01
	P3 (Dyke)	<0,01	<0,01	<0,01	<0,01	<0,01	<0,01	<0,01	<0,01	<0,01	<0,01	<0,01	<0,01	<0,01	<0,01	<0,01	<0,01
	P4 (Right bank)	<0,01	<0,01	<0,01	<0,01	<0,01	<0,01	<0,01	<0,01	<0,01	<0,01	<0,01	<0,01	<0,01	<0,01	<0,01	<0,01

Legend: Lin: Lindane; Aid: Aldrin; **Met:** Methoxychlore; **Mir:** Mirex; **Diel**: Dieldrin; **Hep:** Heptachlore; **Chi**: Chlordimeform; **Chio**: Chlorothalonil; **Meth**: Methazochlore; **Pre**: Pretilachlo

		SYNTHETIC PYRETHROIDS								ORGANOPHOSPHOROUS COMPOUNDS						
	Parameters	**Cpmth**	**Dmeth**	**Lambda-**	**Tmth**	**Prnth**	**Alpha-**	**Cfth**	**Bfth**	**Dzn**	**Dmth**	**Mvp**	**Hptpho**	**Mnctph**	**Ethppho**	**AzphosEt**
	Units	**mg/L**	**mg/L**	**mg/L**	**mg/L**	**mg/L**	**mg/L**	**mg/**	**mg/**	**mg/**	**mg/L**	**mg/**	**mg/L**	**mg/L**	**mg/L**	**mg/L**
	PI (Rive	<0,01	<0,01	<0,01	<0,01	<0,01	<0,01	<0,0	<0,0	<0,0	<0,01	<0,0	<0,01	<0,01	<0,01	<0,01
Codes samples	*P2 (Dyke)*	<0,01	<0,01	<0,01	<0,01	<0,01	<0,01	<0,0	<0,0	<0,0	<0,01	<0,0	<0,01	<0,01	<0,01	<0,01
	P3 (Dyke)	<0,01	<0,01	<0,01	<0,01	<0,01	<0,01	<0,0	<0,0	<0,0	<0,01	<0,0	<0,01	<0,01	<0,01	<0,01
	P4 (Right bank)	<0,01	<0,01	<0,01	<0,01	<0,01	<0,01	<0,0	<0,0	<0,0	<0,01	<0,0	<0,01	<0,01	<0,01	<0,01

Legend: Cpiiithr: Cypermethrin; **Dmethr:** Deltamethrin; **Lambda-Chthr**: Lambda-Cyhalothrin; **Tmthr:** Tetramethrin; **Pmthr:** Permethrin; **Alpha-Cypmthr:** Alpha-Cypermethrin; **Cfthr: Cyfluthrin; Bfth**: Bifenthrin; **Dzn**: Diazinon; **Dmtha**: Dimethoate; **Mvp:** Mevinphos; **Hptphos:** Heptenophos; **Mnctphos:** Monocrotophos; **Ethpphos:** Ethoprophos; **AzphosEth:** Azinphos Ethyl

		CARBAMATES AND OTHERS											
	Parameters	**Qtzen**	**Imaz-**	**Mthmyl**	**Prpcur**	**Dfbenzam**	**Trdmef-**	**Penco-**	**Propico-**	**Azoxystrb-**	**Simazine**	**Benalax-**	**Carbof-**
	Units	**mg/L**	**mg/L**	**mg/L**	**mg/L**	**mg/L**	**mg/L**	**mg/L**	**mg/L**	**mg/L**	**mg/L**	**mg/L**	**mg/L**
	PI (Rive Gauche)	<0,01	<0,01	<0,01	<0,01	<0,01	<0,01	<0,01	<0,01	<0,01	<0,01	<0,01	<0,01
	P2 (Dyke)	<0.01	<0.01	<0.01	<0.01	<0.01	<0.01	<0.01	<0.01	<0.01	<0.01	<0.01	<0.01
Sample codes	*P3 (Dyke)*	<0,01	<0,01	<0,01	<0,01	<0,01	<0,01	<0,01	<0,01	<0,01	<0,01	<0,01	<0,01
	P4 (Right bank)	<0,01	<0,01	<0,01	<0,01	<0,01	<0,01	<0,01	<0,01	<0,01	<0,01	<0,01	<0,01

Legend ; Qtzen : Quintozene ; **Imaz-** : Imazalil ; **Mthmyl** : Methomyl ; **Prpcur** : Prooposcur ; **Dfbenzam** : Diflubenzamide ; **Trdmef-** : Triadimefon ; **Penco-** : Penconazole ; **Propico-** : Propiconazole ; **Azoxystrb-** : Azoxystrobine ; **Simazine** : Simazine ; **Benalax-** : Benalaxyl ; **Carbof-** : Carbofuran

NB: - LQ= Limit of quantification of pesticides is 0.0100

All four water samples analysed showed no residues of the pesticides tested (Results<0.0100)

Categories	Parameters	Description/interest
Conventional physico-chemical parameters/ In-situ measurements	Temperature	Influences chemical equilibrium,
	pH	Measures acidity; distinction between acidic and basic water; linked the nature of the aquifer and environmental conditions
	Conductivity	Indicates the level mineralisation of the water
	Dissolved oxygen and saturation percentage	Determines the biological activity of the environment and its pollution status
	Turbidity	Provides information on water clarity
Organoleptic parameters	Colour	Tells you the colour of the water
	Odour	Tells you what the water smells like
Major elements, trace elements	Potassium	Measures potassium content; indicator of mineralisation
	Sodium	Measures sodium content; indicator of leaching from aquifers containing silicates or sodium chloride
	Chlorides	Measures the content of these elements and provides information on the nature of the aquifer and water pollution
	Nitrates	
	Nitrites	
	Ortho phosphates	
	Sulphates	

	Free cyanides	Measures free cyanide content; provides information on contamination of industrial or commercial wastewater and mining effluent
	Total cyanides	Measures total cyanide content; provides information on contamination of industrial or commercial wastewater and mining effluent
	Copper	Measures copper content; Present in surface water or groundwater; its source is rock or industrial or mining effluents.
	Zinc	Measures Zinc content; Present in surface water or groundwater; its source is rock or industrial or mining effluents.
	Manganese	Measures the Manganese content; its source is the rock or pollution caused by human activities
Organic pollution parameters	COD	Measures the oxygen content required to chemically oxidise organic compounds; Used to assess the quantity of biodegradable and non-biodegradable organic pollutants in surface water.
	BOD5	Measures the oxygen content required to biologically oxidise organic compounds; Enables the biodegradable fraction of the carbonaceous pollutant load in surface water to be assessed within 5 days.
	Oxidability with KMnO	Measures all oxidisable compounds (C, P, N, etc.)

Parameters analysed	Units	Methods	Standards	Equipment used	Detection limits
Temperature	°C	Electrochemical	NF EN 27888	Conductivity meter 3210	0.0 à 60.0
pH		Electrochemical	NF T 90-008	Hanna HI 98128	-2.00 à 16.00
Conductivity	pS/cm	Electrochemical	NF EN 27888	Conductivity meter 3210	0.0 à 3999
Turbidity	NTU	Nephelometry	NF EN ISO 7027	Wag-WT 3020 turbidimeter	0.02
Dissolved oxygen	mg/l	Electrochemical	-	Multimeter 3430 SET F	-
Chloride	mg/l	Chromatography	HACH 8048 method	Ion chromatograph	0.02
Nitrate	mg/l	Chromatography	HACH Method 8114	Ion chromatograph	0.3
Phosphate	mg/l	Chromatography	HACH Method 8008	Ion chromatograph	0.02
Sulphate	mg/l	Chromatography	HACH 8051 method	Ion chromatograph	2
Sodium	mg/l	Atomic absorption	FD T90-112	AAS	0.5
Potassium	mg/l	Atomic absorption	FD T90-112	AAS	0.5
Copper (Cu)	mg/l	Atomic absorption	FD T90-112	AAS	0.05
Manganese (Mη)	mg/L	Atomic absorption	FD T90-112	AAS	0.05
Zinc (Zn)	mg/l	Atomic absorption	FD T90-112	AAS	0.05
Cyanides	mg/l N-	Spectrometry (Pyrazalone-pyridine method)	-	DR3900	0,002
Nitrite	mg/l	Spectrometry (Diazotation method)	-	DR3900	0,002
Pesticides	mg/l	Gas chromatographic method after liquid-liquid extraction.	NF EN ISO 6468, NF EN ISO 11369	GC	0.0100

PART II

BURKINA FASO

Unity - Progress - Justice

MINISTERE DE L'ENVIRONNEMENT,
DE L'EAU ET DE L'ASSAINISSEMENT

SECRETARIAT GENERAL

MINISTRY OF THE ENVIRONMENT,
WATER AND SANITATION

GENERAL SECRETARIAT

REPORT Summary

Investigation into water pollution at a well in Dédougou, MOUHOUN BOUCLE REGION

November 2023

INTRODUCTION

The galloping demography of African towns in general, and Dédougou in particular, is causing problems with drinking water supplies. This situation is forcing the inhabitants to use well water, which has no information on its bacteriological and physico-chemical quality. However, wells, the ancestral symbol of access to this precious substance in rural areas, are now facing an invisible and devastating threat. Pollution of water resources is characterised by the presence of micro-organisms, chemical substances and industrial waste. The pollution of water resources has worsened and is becoming more and more recurrent in Burkina Faso with the use of chemicals such as cyanide and mercury at mining and gold-panning sites, and is worsening with the rampant insecurity in certain regions, including the Boucle du Mouhoun.

As indicated, on Monday 06 November 2023, following the poisoning of certain members a family who had consumed water from a traditional well in the Boucle du Mouhoun region, more precisely in the village of Passakongo located about ten (10) kilometres from Dédougou, the General Directorate of Water Resources was called upon to carry out investigations in order to clarify the situation of the population. On the instructions of the Secretary General of the Ministry of the Environment, Water and Sanitation, a joint mission led by agents from the Directorate General for Water Resources (DGRE) and the Directorate General for Environmental Preservation (DGPE) travelled to Dédougou on Wednesday 08 November 2023 and then to the village where the incident took place. Water samples were also taken by the water police on 15 February 2024 for retrocontrol.

The overall aim of the assignment was to investigate the phenomenon of pollution of the resource and define measures to protect people and property.

More specifically, this :

- Meet the authorities at regional level and inform them of the mission's objective;
- Talk to the villagers to understand the facts of the incident;
- Take water samples for analysis to identify the source of the pollution;
- Formulate recommendations for sustainable quality management of water resources.

The expected results were :

- The regional authorities are informed of the purpose of the mission;
- The people of the village showed an interest in understanding the facts of the incident;
- The water samples taken are analysed and the pollution index is plotted;
- Recommendations are made for the sustainable qualitative management of water resources.

1. OVERVIEW OF THE MISSION

The team **first** visited the Boucle du Mouhoun Regional Water and Sanitation Department (DREA-BM) in Dédougou, where it met with Mr Koura Bakary, the officer in charge of water policing at the department, to discuss the purpose of the mission and hear his version of events.

Our discussions enabled us to gather the following information:

- The name of the village where the incident took place is Passakongo, around 9 kilometres from the centre of Dédougou.
- A joint mission (DREA-BM) from the Mouhoun Water Agency (AEM) visited the site on the evening of Monday 06 November 2023, to observe the animal deaths and take water samples for possible analysis;
- He contacted the person from the Water Agency responsible for reporting on their joint mission to join us at the DREA-BM so that we could go together to the site of the incident;
- The report from the Dédougou Health District was provided to us for further information, pending the arrival of the agent.
- The information contained in this report is shown in Table I :

Table 4: Spot report (SpotRep) of drinking water poisoning from a well of two members a family in the Passakongo village in the Passakongo health area, Dédougou health district.

ELEMENTS	DESCRIPTION
Date of writing	06 November 2023 at 11 hours 56 minutes
Title or name of incident :	Death 6 pets Consultation of 02 people from the same family at the CSPS in Passakongo
Sources of information	Head Nurse of the CSPS of Passakongo
Statement of facts:	This morning a woman and her child came to the CSPS. They all had the same

	symptoms, namely vomiting, stomach ache and diarrhoea. On questioning, they were said to have drawn and drunk water from the well where they usually get their drinking water. 5 goats and a donkey died after drinking from the same spring this morning. The people accompanying the patient and those interviewed identified the well water as the source of the poisoning.
ELEMENTS	DESCRIPTION
Actions :	- Informing management - Informing village authorities - Involvement of communities in monitoring to prevent consumption of well water - District EIR activated for investigation in the village today - Symptomatic management of patients by staff through the administration of antipyretics and rehydration. - The closure of the offending well - Seizure of dead animals to prevent their consumption by the public - Setting up a surveillance team to prevent consumption of the well water - The process officers were advised to draw up a line list and to give advice to the escorts and the local population.
Information control :	This situation is reported by the DRS for Health in the Boucle du Mouhoun region.
Contact point for the report	Dr Nathalie W. OUADEBA. Chief Medical Officer for the Dédougou health district Tel: (00226) 70 45 36 37

Shortly afterwards, the AEM agent, Mr DIBLONI Sami Edgard, in charge of reporting on their joint mission, arrived.

After the courtesy greetings, he gave us a report on their mission.

The following information has been selected:

- Following the fainting of a woman and her child after drinking water from a traditional well, a joint team from the Mouhoun Water Agency and the Boucle du Mouhoun Regional Water and Sanitation Department rushed to the scene of the tragedy to gather information;
- A team from the National Public Health Laboratory will also be visiting;
- The report describes case of a woman and her child who drank well water that turned out to be undrinkable. They fainted in front of the well and were found by her husband, who took them to a health centre. Their condition remained critical as we left the scene **on 06 November 2023**.
- It should also be noted that the animals that consumed the same water all died three (03) to five (05) minutes later, according to witnesses. They were animals belonging to DAKUO families living near the well in question.
- The joint AEM and DREA-BM mission was able to take samples of water from the well for analysis. A total of five (05) cans containing water from the well (three for the agency and two for the regional office) were taken, taking into account the safety measures required for our well-being.
- The village of Passakongo, where the incident took place, is in a red zone.
- Alert security before any movement in the area.

The same information was obtained from the Regional Director of the Environment and the Chief Medical Officer of the District of Dédougou, whom we met today.

Secondly, following the interviews, the mission moved to the site of the incident, accompanied by KOURA Bakary and DIBLONI Sami Edgard, to make observations and take water samples for analysis. We escorted by two National Police officers. We were also assisted by an officer from the Centre d'Information Sanitaire et de Surveillance Epidémiologique (Centre for Health Information and Epidemiological Surveillance) and the Major of the Passakongo CSPS.
Once on site, we found the CVD of the village of Passakongo. Discussions with the Chief enabled us to gather the same information as we had obtained from the AEM and DREA-BM agents and from the health report.
Observations and hearings in the field provided the following information:
- The well has been in existence for almost six (06) years;
- It is located between the concessions ;
- This is a traditional well with a curbstone and no cover;
- When we arrived, the well in question had already been sealed with sheet metal, as shown **in Appendix 2;**
- As it approached, it smelt of cyanide, which became more pronounced when the metal sheet was removed for a closer look;
- According to Mr KOURA, in charge of water policing, it was the smell of cyanide;
- Some members of the team began to suffer from headaches;
- When asked if they had not noticed a suspect in the vicinity of the well, the CVD's response was that on Sunday 05 November 2023, there had been up to two funerals in the village, so it was the following day that the incident was noticed;
- The exact number of dead animals is two donkeys and five goats (see Appendix 2);
- Four people were affected: a woman and her young son, who were admitted to hospital on the morning of 06 November, her daughter, who was able to vomit contaminated water on the spot, and a man, who also escaped;
- In all the years that the well has been dug, there has never been a case like this;
- The inhabitants and their animals drink the water from this well;
- For the number of water points in the area, we: At 46 metres, in the same
A third well is located one hundred (106) metres north of the first one. There are at least three boreholes in the vicinity, the nearest of which is about 150 metres away but inaccessible;
- Reported on site on Wednesday 08 November: the woman, her young son and her daughter, who were poisoned by well water, are doing well;
- The dead animals were incinerated;
- There are no gold-panning sites in the vicinity of the village;
- As of 10 November, those poisoned were doing well, and the second well has since been closed, they said.
After making the usual observations, the team took water samples and *took in situ* measurements at the contaminated well, two other wells and the school borehole. Their GPS coordinates were taken.
- The appearance of the water from the offending well was pink, as was the water from the second well. Only the water from the third well was clear and limpid (**see Appendix II**);
- According to witnesses, the contaminated appearance of the well water is explained by the fact that the water was stirred **on 07 November 2023** in an attempt to find signs of the pollution in question, but to no avail;

- So the agitation of the water led to chemical reactions that gave the water in the well in question its pink appearance, otherwise it was grey in colour from 06 November, they said (**see Appendix 2**);
- The water samples taken are kept cool in coolers. They are then sent to the DGRE's Raw Water Analysis Laboratory (to analyse additional parameters in some cases, and to the Environmental Quality Analysis Laboratory others).

In order to monitor the pollution of the well, water samples were taken on 15 February 2024 from the three (03) wells in the contamination zone for back-checking. This sampling was facilitated by Mr KOURA Bakary of the water police at the Mouhoun Regional Water and Sanitation Department in Dédougou and the CVD of the village of Passakongo.

2. RESULTS AND INTERPRETATION

2.1. Analysis equipment and methods used

The equipment and methods used to analyse the parameters studied are listed in the table in **Appendices III and IV**.

2.2. Presentation of analysis results

Given that water from traditional wells is generally undrinkable, and that witnesses to the incident suspected cyanide poisoning, this enabled us to direct our analyses. The essential parameters were selected.

Tables II and III show the analytical results of the water samples taken for the two sampling campaigns (first and second sampling).

Table 5: Analysis results for the first sample (November, 2023)

EchantiHons		*Analysis results*														
Sample code	*Geographical location*	*T (°C)*	*pH*	*Conductivity f/jS/cm)*	*Turbidite (NTU)*	*NO¿ (mg/D*	*Na (mg/l)*	*K (mg/l)*	*Cu (mg/l)*	*Mn (mg/l)*	*Fe (mg/l)*	*Zn (mg/l)*	*CN (mg/l)*	*(mg/l)*	*N03- (mg/L)*	*W' (mg/L)*
Pl(Puitsl)	I2⁰3Γ24,2 "N OD3°25'I5.2 "W	32,4	9,53	3140	313,7	3,5	559	ILI	0,075	0,151	0,44	2,71	2,88	2,41	3,4	5
P2(Well2)	I2⁰3Γ25,2 "N 0D3°25'IG,3 "W	32,1	5,94	55,2	143	<0,02	3,318	5,51	<0,05	<0,05	1,31	<0,05	0,35	1,83	3,4	12
P3(Well3)	I2°3I'27,G "N □□3°25'14,3 "W	33	4,45	72	17,83	<0,02	3,58	7,45	<0,05	<0,05	0,12	<0,05	0,002	0,28	0,2	<2
P4 (èco fe borehole)	12°31'34,3 "N □□3°25'13,7 "W	32,7	5,32	55	10,43	0,002	2,011	8,588	<0,05	<0,05	0,3	<0,05	0,018	<0,11	28,2	<2
Decree no. 201 185/PRES/PI Article? no quelites des petebiliseble! pollution	1- 1/MEE rmes of нıx per degree of	IB- 40	5,5 è B,5	1000		0,2	30	30	0,05	0,05	0,3	0,5	0,05	1,5	50	150
Arrêté Conjoint 00019-2005/MAHRH/MS portent definition des			<B		5	3	200		1	0,1	0,3	3	0,07	1,5	50	250

normes de potebilité de Геей																

In red: value exceeding the standard, Good: normative value, Other: value below the standard.

Table 6: Analysis results for the second sample (February, 2024)

EchantiHons		*Analysis results*														
Coda samples	*Geographical location*	*T (°C)*	*pH*	*Conductivity (µ S/cm)*	*Torbidità (NTU)*	*HO/ (mg/i)*	*Ha (mg/D*	*K (mg/1)*	*Cu (mg/D*	*Mn (mg/D*	*Fa (mg/D*	*Zn (mg/D*	*C№ (mg/D*	*NH4+ (mg/D*	*NI13-* (mg/L)	*W' (mg/L)*
PI(Puitsl)	I2^{O}3Г24,2 "N OD3°25'I5.2 "W	30,4	7,4G	428	72,D8	0,15	4,75	0,55	□,□72	0,097	0,66	□,II8	□,□17	6,97	1,4	1,2
P2(Well2)	I2^{O}3Г25,2 "N 0D3°25'IG,3 "W	31,1	6,20	58,5	33,74	D,DDG	0,55	<D,5	<D,D5	0,091	□,□78	<D,D5	0,010	1,36	0,7	0,2
P3(Well3)	I2°3I'27,G "N □□3°25'14,3 "W	30	5,90	GD	22,D8	0,81	0,5	<D,5	<D,D5	0,093	0,20	0,67	□,□□5	0,81	<D,2	<D,2
Decree n°201 185/PRES/PI Article? no quelites des petebiliseble! pollution	1- 1/MEE rmes of них per degree of	IB- 40	5.5 è B.5	1000		0,2	30	30	0,05	0,05	0,3	0,5	0,05	1,5	50	150
Arrêté Conjoint 00019- 2005/MAHRH/MS portent definition des normes de potebilité de Геей			<B		5	3	200		1	0,1	0,3	3	0,07	1,5	50	250

In red: value exceeding the standard, Good: normative value, Other: value below the standard.

2.3. Interpretation of analysis results

For all the parameters studied in this study, it is important to note that several values exceed the potability standards specified in the Joint Order for the samples from well P1 in question (**see table 2**).

2.3.1. Temperature

The temperature (T) of water affects its density and viscosity, the solubility of oxygen and the speed of chemical and biochemical reactions.

The values observed (32.1°C to 33°C) are within the acceptable temperature range for drinking water (18-40°C), which is in line with general standards. With specific regard to the well in question, the two samples have different temperatures (32.4°C for the first and 30.4°C for the second), but both are within the normal range for drinking water (**tables 2 and 3**). Minor variations do not appear to be of concern for water potability (**Smith et al., 2017).**

2.3.2. pH

The pH is the cologarithm of the concentration of hydrogen ions in water: 0 to 7 is acidic and 7 to 14 is alkaline.

The pH varied from 4.45 to 9.53 during the first sampling. These deviations from standards (5.5 to 8.5) suggest a significant divergence. As for the well in question, the pH is relatively high on the first sample (9.53) but falls within an acceptable range (7.46) on the second sample.

More acidic or basic values could impact water quality and affect potability, potentially through chemical reactions with dissolved elements **(Vengosh et al., 2018).**

2.3.3. Conductivity

Electrical conductivity (EC) is a measure of the ability of an aqueous solution to conduct an electric current, which depends on the concentration of ions in solution. It is sensitive to variations in dissolved matter, mainly mineral salts. Conductivity also increases with water temperature.

Most of the observed values are well below the maximum limit of 1000 µS/cm, which is generally acceptable for drinking water. This indicates a low concentration of dissolved ions **(Huang et al., 2019).** However, in the case of well P1, the conductivity value was too high (3140 µS/cm) during the first sampling. This value fell considerably during the second sampling (428 µS/cm. This could be explained by the chemical and biochemical reactions generated with the toxic compound due to the agitation of the water in the well in question, which made the water hard during the first sampling.

2.3.4. Turbidity

The turbidity of the water depends on the nature of the terrain it crosses, the season, the rainfall, the water flow regime and the nature of the discharges. It can be assessed by the transparency of the water.

The turbidity of the samples ranged from 10.43 NTU to 313.7 NTU for the first sample. At the well in question, the first sample (313.7 NTU) was much more turbid than the second sample (72.08 NTU). These values exceed the limit of 5 NTU set by the Joint Order. This may be due to the fact that some of the wells in our study do not have coping stones and are not all covered. As a result, run-off water and/or dust can easily enter these wells.

2.3.5. Metallic elements (copper, iron, zinc, manganese) :

The concentrations of these metals in the samples vary. For copper, the limit is 0.05 mg/l according to the decree. Samples from Puits 1 and Puits 3 exceed this limit. For manganese, the standard is 0.1 mg/l. Only Puits 1 exceeds this value. For iron, the standard is 0.3 mg/l, and Puits 1 exceeds this limit. For zinc, the limit is 0.07 mg/l, and only Puits 1 exceeds this limit (**Table 2**).

Most of the samples met the standards, except for iron in one case. An excess of iron can alter the taste and colour of the water but does not necessarily affect its potability **(Soleimani et al., 2016).**

Concentrations of trace metals vary considerably between the two samples. In some cases, the concentrations in the first sample exceed the standards set by the decree, while in others they are lower.

2.3.6. Other contaminants (Nitrites, Ammonium, Cyanides, Nitrates and Sulphates) :

Concentrations vary for each sample. It is important to note that several values exceed the potability standards specified in the Joint Order for the samples from wells P1 and P2, particularly with regard to cyanide.

For the first sampling, cyanide concentrations in samples from Puits 1, Puits 2 and Puits 3 ranged from 0.002 to 2.88 mg/L. According to the decree, the standard for drinking water is 0.05 mg/l. Similarly, the World Health Organisation (WHO) recommends that cyanide concentrations should not exceed 0.07 milligrams per litre (mg/L) or 70 µg/L. Samples from Wells 1 and 2 exceeded this standard, while those from Well 3 and the borehole were below it. The water from these two wells was therefore unfit for human and animal consumption. This could be the cause of the poisoning observed in patients and animals.

For the second sampling, all three (03) wells showed cyanide concentrations below the standard, and therefore compliant (**table 3**). Well 1 showed a much lower concentration of cyanide (0.017 mg/l), which is well below the authorised standard for drinking water. This indicates a significant improvement in water quality compared with the first sample, but it is important to note that even low concentrations of cyanide can be dangerous to human health in the long term.

Cyanides are extremely toxic to animals. A concentration of 200 microgrammes per litre ^g/L) in water can be dangerous or even fatal for animals, depending on the length of exposure and the quantity consumed.

Cyanide toxicity manifests itself differently depending on the individual or the receiving environment, the mode of exposure and the form of the cyanide, which explains the different reactions of intoxicated patients following consumption of this contaminated well water. Exposure to cyanide can occur by inhalation or through the skin. Cyanide does not bio-accumulate in living organisms. It acts directly as a poison. Hydrogen cyanide (HCN) is the most toxic of all cyanide compounds. For complex cyanides, toxicity is due to the ease with which the free cyanide dissociates from the complex.

The main route of exposure for humans is inhalation. Since cyanide does not bioaccumulate in living organisms, its most serious toxic effects occur within organisms and are due to the combination of CN- with the iron in haemoglobin, preventing oxygenation of the blood **(MCCARTY, 2002)**. Once released into the environment, cyanide behaves in several ways: complexation, adsorption, precipitation, volatilisation and biodegradation. Although cyanide reacts easily in the environment to form complexes and salts of varying stability or degrades, it is highly toxic to living organisms at very low concentrations (**B. KONÉ, 2015).**

Given that the village is located in an insecure area, the high concentration of cyanides in the well water in question could be due partly to a case of crime, poisoning or settling of scores, and partly to the infiltration of surface water and the lack maintenance of these wells. Collecting water from a well that has been abandoned on the ground is a further source of well water contamination.

The protection of groundwater depends first and foremost on the proper maintenance of boreholes, which are liable to cause direct contamination, and on the proper management of water abstraction through them. But all the players in the catchment area are concerned, and must be involved in the overall management of water resources. Complemented by specific regulations, this overall management helps to preserve groundwater.

Promoting the potabilisation of water at home and raising awareness of appropriate practices is essential to guarantee food safety and health for all. There are a number of measures that can be taken to ensure that water in the home is safe to drink. These can include using water filters to remove impurities and contaminants, boiling the water to kill bacteria and parasites, or using disinfectant chemicals such as bleach or activated carbon or lime.

RECOMMENDATIONS

The following recommendations (**Table 4**) were made following the investigations for sustainable quality management of water resources.

Table 7: Recommendations for quality management of water resources

N°	Recommendations	Deadline for	Structure of d'iivre

		implementation	
1	Ensure that the offending well is closed, as well as the second P2 well in the area close to and in the same direction as the contaminated well, to avoid any poisoning between now and the rains.	New order	CVD, DREA, AEM
2	Open the wells in question as soon as the rains fall to encourage water purification.	June 2024 possible	CVD, Water Police
3	Install more drinking water collection facilities (boreholes, water supply systems, etc.) accessible to all in the area, for the well-being of the population.	Continue	MEEA
4	To educate the population and raise awareness of good public hygiene practices around water points in order to prevent water-borne diseases.	Continue	MEEA, Water Agencies, DREA, Ministry of Health
5	Strengthening water quality control and monitoring systems (water quality monitoring networks, strengthening the technical facilities of DGRE and DGPE laboratories).	Continue	MEEA, DGRE, DGPE
6	Establish a mechanism for rigorous monitoring and control of the movement of cyanide from checkpoints at the country's borders	Continue	DGPE, ANEVE, BUMIGEG, Security, Customs, LNSP(ANSSEAT)

CONCLUSION

The aim of the mission was to investigate the phenomenon of well water pollution in the village of Passakongo, around 10 kilometres from Dédougou, and to define measures to protect people and property.

The study carried out on the water from three (03) wells and the borehole at the village school assessed its physico-chemical quality. Well water is still an important source of water supply in Dédougou. It is used for a number of purposes, especially food.

The first sample suggests severe cyanide contamination of the water, while the second sample shows an improvement but still requires special attention in terms of drinking water quality. Government regulations must be followed closely to ensure the safety and public health of the water supply.

In order to prevent water-borne diseases, it is essential to increase the number of drinking water collection points and to educate and raise awareness among the population about good public hygiene practices around water sources ( ).

BIBLIOGRAPHICAL REFERENCES

- Joint Order 00019-2005/MAHRH/MS defining drinking water standards ;
- B. KONÉ, É. K. YAO, B. SILUÉ, G. CISSÉ, N. SORO, Approvisionnement en eau potable, qualité de la ressource et risques sanitaires associés à Korhogo (Nord-Côte d'Ivoire), Environnement, Risques & Santé, 14 (2015);
- C. N. SAWYER, P. L. MCCARTY, Chemistry for sanitary engineers, (1967) [20] - J. P. CHIPPAUX, S. HOUSSIER, P. GROSS, C. BOUVIER, F. BRISSAUD, Etude de la pollution de l'eau souterraine de la ville de Niamey, Niger, Bull Soc Pathol Exot, 94 (2002) 119-123 [21] ;
- Decree n°2001-185/PRES/PM/MEE, Article 7 Water quality standards by degree of pollution ;
- Huang, H. (2019). Conductivity as an Indicator of Water Quality;
- WHO drinking water standards (2017) ;
- Soleimani, H. (2016). Iron in Drinking Water: Is it Safe;
- Smith, J. (2017). Water Quality Standards and Their Role in Water Quality Management;
- Vengosh, A. (2018). The Chemistry of Water;

APPENDIX 1: LIST OF RESOURCE PERSONS MET IN THE LOCALITY

Surname and first name(s)	Structure	Contacts	
KOURA Bakary	Boucle du Mouhoun Regional Water and Sanitation Department	70 75 16 55	
DIBLONI Sami Edgard	Mouhoun Water Agency	Tel	70 68 77 13
DAKYO Matthieu	Village chief (CVD)	Tel	71 03 51 58
Warrant Officer PARE Manassé	National Police	Tel	56 13 16 66
Warrant Officer SOALA Roland	National Police	Tel	76 45 16 32
Dr Nathalie W. OUADEBA.	Dédougou Health District	Tel	70 45 36 37
COULIBALY Ali	Boucle du Mouhoun Regional Environment Department	Tel 768	70163614/ 07494
KIENOU Ilarian	Centre for Health Information and Epidemiological Surveillance	Tel : 70 61 11 30	
SOMPOUGDOU Paul	Passakongo CSPS	Tel : 70 52 96 81	

APPENDIX 2: ILLUSTRATIVE PHOTOS, DGRE, NOVEMBER 2023

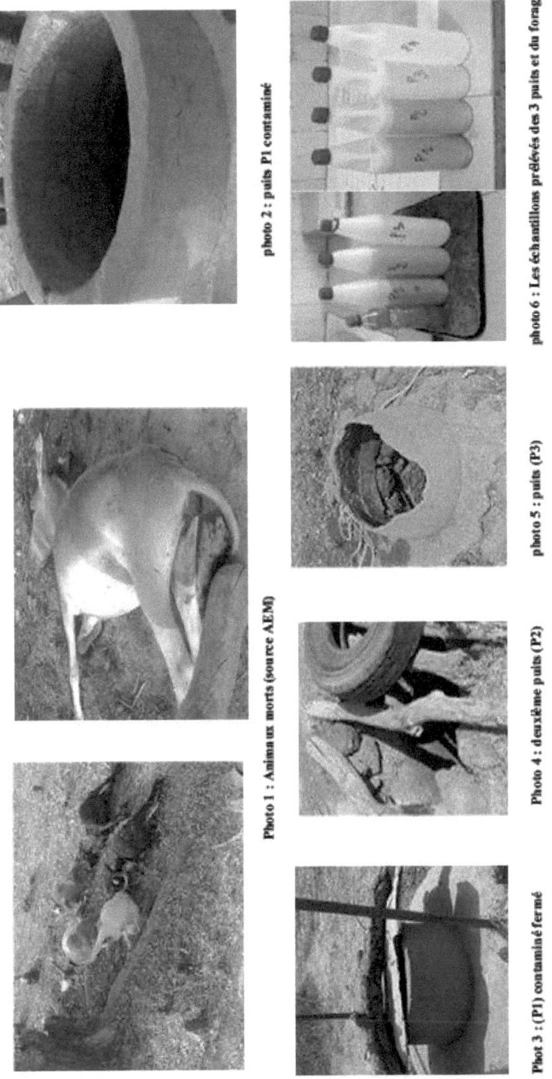

Photo 1 : Animaux morts (source AEM) photo 2 : puits P1 contaminé Phot 3 : (P1) contaminé fermé Photo 4 : deuxième puits (P2) photo 5 : puits (P3) photo 6 : Les échantillons prélevés des 3 puits et du forage

Photo 1: Dead animals (source AEM) photo 2: contaminated well P1 Photo 3: (P1) contaminated closed Photo 4: second well (P2) photo 5: well (P3) photo 6: Samples taken from the 3 wells and the borehole

APPENDIX 3: TABLE I: DEFINITION OF SOME OF THE PARAMETERS STUDIED

Categories	Parameters	□escription/interest
Classical physico-chemical parameters/ In-situ measurements	Temperature	Influences chemical equilibrium.
	pH	Measures acidity; distinction between acidic and basic water; linked the nature of the aquifer and environmental conditions
	Conductivity	Indicates the level mineralisation of the water
	Dissolved oxygen and percentage saturation	Determines the biological activity of the environment and its pollution status
	Turbidite	Provides information on water clarity
Organoleptic parameters	Colour	Tells you the colour of the water
	□deur	Find out more about Todeur de l'eau
Major elements, metallic trace elements	Potassium	Measures potassium content; indicator of mineralisation
	Sodium	Measures sodium content; indicator of l ixiviati on of aquifers containing silicates or sodium chloride
	Chlorides	Measures the content of these elements and provides information on the nature of the aquifer and water pollution
	Nitrates	
	Nitrites	
	□rtho phosphates	
	Sulphates	
Categories	**Parameters**	**Description/interest**
	Free cyanurES	Measures the content of free cyanides; provides information on contamination of industrial or artisanal wastewater and mining effluents
	Total CyanuPES	Measures total cyanide content; provides information on contamination of industrial or commercial wastewater and mining effluent
	CuivPE	Measures copper content; Present in surface water or in fresh water; its source is rock or effluent from industry, mining, etc.
	Zinc	Measures zinc content; Present in surface water or groundwater; source is rock or industrial or mining effluents.
	Manganese	Measures manganese content; its source is rock or pollution caused by human activity

APPENDIX 4: TABLE I: SOME OF THE ANALYSIS EQUIPMENT AND METHODS USED

Analysis parameters	Units	Methods	Standards	Equipment used	Detection limits
Temperature	°C	Elect rochimigue	NF EN 27888	ConductimétrE 32ID	O.OàGO.O
pH		Elect rochimigue	NET 9D-DD8	Hanna HI 98128	-2.00ÒIG.00
Conductivity	pS/cm	Elect rochimigue	NF EN 27888	ConductimétrE 32ID	0.0 à 3999
Turbidite	NTU	Nephelometrigue	NFENISD7D27	Turbidimetre 355 IR	0.02
Chloride	mg/1	Chromatography	HACH Method 8D48	ChromatographE Ionigue 1C Mctrohm	0.02
Nitrate	mg/1	Chromatography	HACH Method 8114	ChromatographE Ionigue 1C Mctrohm	0.3
Phosphate	mg/1	Chromatography	HACH Method 8DD8	ChromatographE Ionigue 1C Mctrohm	0.02
Sulphate	mg/1	Chromatography	HACH 8D5I method	ChromatographE Ionigue 1C Mctrohm	2
Sodium	mg/1	Atomic absorption	FD T9D-II2	SAA Perkin Elmer Pin AAClE 9DDT	0.5
Potassium	mg/1	Atomic absorption	FD T9D-II2	SAA Perkin Elmer Pin AAClE 9DDT	0.5
Copper (Cu)	mg/1	Atomic absorption	FD T9D-II2	SAA Perkin Elmer Pin AAClE 9DDT	0.05
Manganese (Mn)	mg/L	Atomic absorption	FD T9D-II2	SAA Perkin Elmer Pin AAClE 9DDT	0.05
Zinc (Zn)	mg/1	Atomic absorption	FD T9D-II2	SAA Perkin Elmer Pin AAClE 9DDT	0.05
Cyanides	mg/1 N-	Spectrophotometry (Pyrazalone-pyridine method)	-	DR39DD	0,002
Nitrite	mg/1	Spectrophotometric (Diazotati on method)	-	DR39DD	0,002

PART III

BURKINA FASO

Unité - Progrès - Justice

MINISTERE DE L'ENVIRONNEMENT,
DE L'EAU ET DE L'ASSAINISSEMENT

SECRETARIAT GENERAL

BURKINA FASO

Unity - Progress - Justice

MINISTRY OF THE ENVIRONMENT,
WATER AND SANITATION

GENERAL SECRETARIAT

MISSION REPORT

Inspection and sampling of water of the River Poni following illegal and processing of gold by the mining company LSM Gold Corporation in the Commune of Gbomblora, Province of Poni, South West Region

January 2024

INTRODUCTION

Pollution of water resources is characterised by the presence of micro-organisms, chemical substances or industrial waste. This pollution can have various origins (industrial pollution, agricultural pollution, domestic pollution, accidental pollution). The pollution of water resources has worsened and is becoming increasingly recurrent in Burkina Faso with the use of chemicals such as cyanides at mining and gold-panning sites.

Following a complaint from the people of Ouadaradouo and surrounding villages in the commune of Gbomblora, Poni province, South-West region, about water pollution in the river caused by mining activities, the Directorate-General for Water Resources was called in to take samples to establish the position of the various parties.

MISSION OBJECTIVES

Overall objective

Carry out another site visit, talk to local people and take samples for laboratory analysis.

Specific objectives

The specific objectives of the assignment are :

- Meet and discuss with local and regional authorities to inform them of the objectives of the mission;
- exchange with local people;
- carry out in-situ analyses and take water samples for in-depth laboratory analysis.

EXPECTED RESULTS :

The expected results at the end of the assignment are :

- regional and local authorities are informed of the purpose of the mission;
- Discussions are held with the public;
- In situ analyses and water samples are for in-depth laboratory analysis.

STATE OF PLAY

The mission was made up of a team from the Ministry of Energy, Mines and Quarries, the Directorate General for Water Resources, the Directorate General for Environmental Preservation, the Poni Provincial Environment Directorate and an officer from the South-West Water Police. The mission took place on Sunday 21 January 2024.

Arriving in Gomblora at 8.40am, the mission met with the Prefect, President of the commune's Special Delegation, during which he explained the background to the events and gave us advice on how to avoid any inconvenience on the ground.

He provided us with the various reports from previous meetings.

We left Gomlora and arrived in Ouadaradouon at 10am to find a barrier had been erected by the local population to prevent access to the site. After talking to them, they said that this mission was one too many and that the missions carried out to resolve the problem up to then had been unsuccessful, and that they didn't want any more missions.

Nevertheless, we were able to access the first site and found that the water in the river was visibly cloudy.

At the work site, the water has been artificially diverted by the mining company. There is an earth dam on the riverbed. The mining company has dug a canal to divert the water from . This canal will not go far, as the same company has used it to create a reservoir (an artificial pond) for processing its ores, thus creating a discontinuity in the flow of water and, as a result, a probable disturbance of the aquatic ecosystem both upstream and downstream. As for the quality of the water, analyses are underway to check whether the water has been polluted.

In fact, mining is even carried out on the minor bed of the watercourse.

As far as the site is concerned, the work seems to have stopped, but as far as the population is concerned, the work continues clandestinely at night. Heavy machinery is still on site, guarded by two security guards day and night.

Once the inspection of this site was complete, we returned to the village (Ouadaradouon) and were then accompanied to the second site at Wakindouon. There was also a strong mobilisation of the population. People from the villages we pass through on our way to the site joined us to get there.

New mining equipment was found along the way, and discussions with the equipment's custodian revealed that it belonged to the same mining company.

When we arrived at the site (Wakindouon), the environmental destruction was the same as at the first site. Water samples were also taken for analysis.

The public is giving the authorities one week to resolve the problem, failing which they will do so in their own way.

It should be noted that we were accompanied by huge crowds at both sites.

On our return from the mission, we gave an update to the PDS of the commune of Gomblora and to the High Commissioner of Poni, who immediately called the promoter to inform him of the mission's presence and of what was being said on the ground about the clandestine pursuit of activities. The latter said that the information was false and that no activities were taking place on the ground.

OBSERVATIONS MADE ON THE SITES

The findings from the field are as follows

- occupation of the river banks right down to the riverbed;
- alluvial soil is taken right down to the riverbed;
- dykes have been built prevent run-off;
- Large, deep excavations 50 metres in diameter and 15 metres deep were certainly made to recover the alluvial soil,
- sludge tending to create a form of wetland at the foot of the ore washing platforms;
- destruction of woody species.

In view of the above, the following regulations have been infringed:

- Law n°002-2001/AN of 8 February 2001 on the orientation of water management through articles 24, 26 and 27 and punished by article 60 et seq;
- Articles 234 and 235 of Law N°003-2011/AN of 05 April 2011 on the BF Forestry Code, the penalties for which are set out in Articles 264 and 265;
- Articles 25 and 71 of Law N°006-2013/AN of 02 April 2013 on the Environmental Code in Burkina Faso, which emphasises the need to obtain prior authorisation after carrying out an environmental and social assessment study; and as a result, punishable under Articles 126 and 135 of the said law, with work being halted in accordance with the terms of Article 103.

ANALYSIS RESULTS

Sample code	T (°C)	pH	Conductivity (pS/cm)	Turbidity (NTU)	Bicarbonates (mg/1	NO_2 (mg/1)	Ca (mg/L	Cu (mg/1)	Mn (mg/1)	Fe (mg/1)	Zn (mg/1)	CN- (mg/1)	NH4+ (mg/1)	BODs (mg/1)	COD (mg/1)
Pl-Sl	27,3	6,67	93,1	1037	41,97	0,02	14,91	0,17	0,48	0,20	0,04	0,003	0,58	405	576
P2-S1	27,5	6,63	100,4	1043	31,11	0,05	15,07	0,18	0,87	0,19	0,04	0,011	2,03	306	422
P3-S1	27,6	7,21	262,0	24,01	64,54	0,03	36,87	0,07	0,09	0,37	0,00	0,008	0,82	103	136
P1-S2	27,8	6,89	119,7	407,1	33,06	0,02	16,99	0,07	0,19	0,10	0,15	0,005	0,72	32	47
P2-S2	27,1	6,90	104,9	778,5	31,72	0,02	15,87	0,10	0,04	0,23	0,00	0,002	0,79	19	28
Drilling (dimora)	24,3	7,11	469,0	1,18	197,15	0,03	70,46	0,07		0,03	0,60	0,002	0,50		

Decree № 2015-1205/PRES-TRANS/PM/MERH/MARHASA/MS/MICA/MME/MIDT/MATD on standards and conditions for wastewater discharges	18-40	6,5 à 9		-				2	5	10		-	-	40	150
Decree №2001-185 PRES/PM/MEE setting standards for the discharge of pollutants into the air, water and soil	-	-	1 000	5		0,2	500	-	-	-	5	1	1	50	150

Site№1: **Ouadaradouon** Site №2 : **Wankidouon**

INTERPRETATION OF RESULTS

Following additional laboratory analyses, the following results were obtained.

Turbidity (NTU) :

Turbidity values varied significantly between samples, ranging from 24.01 NTU to 1043 NTU, above current standards. The high turbidity confirms the significant presence of suspended particles in the water, as observed visually (see illustrative photos).

Cyanide (CN-)

The presence of cyanide was observed in low concentrations in the samples taken, ranging from 0.002 mg/l to 0.011 mg/l. These traces of cyanide could indicate cyanide pollution that has dissipated over time.

Ammonium (NH4+)

Ammonium concentrations vary from 0.5 mg/l to 2.03 mg/l. The presence of ammonium may be linked to the decomposition of organic matter.

Chemical Oxygen Demand (COD)

COD values are quite high, ranging from 28 mg/L to 576 mg/L. This indicates the amount of oxygen required to oxidise the organic compounds present in the water.

5-day Biological Oxygen Demand (BOD5)

BOD5 values range from 19 mg/L to 405 mg/L. This measures the amount of oxygen consumed by micro-organisms over a 5-day period, which is an indicator of the organic load of the water.

Borehole (Ounora): The data for the Ounora borehole show relatively low levels for all parameters measured, suggesting better water quality compared to the sampling points.

In conclusion, the analyses reveal significant variations in water quality between the various sampling points, with high concentrations of certain parameters such as turbidity and chemical and biochemical indices. Immediate steps must be taken to stop gold extraction and processing in the bed of the River Poni and to restore the site to its original state.

RECOMMENDATIONS

In view of the pollution observed (visual, physical, chemical) following the outing and laboratory analyses, we are making the recommendations set out in the table below:

N°	Recommendation	Manager	Other players	Deadline	Comments
1	Ensure strict compliance with works suspension	Governorate		Immediate	Work continues underground

	measures ordered by the Prefect				
2	Transmit the PV de constant and the constituent elements to the Public Prosecutor of the Gaoua TGI (High Court of Gaoua) to initiate legal proceedings for pollution and modification of the water regime against the promoter SOME Benild and the LSM Gold Corporation.	South-West Water Police Department	DREA/SUO DGRE AJE	Immediate	The official report drawn up must be filed with the Public Prosecutor. Also attach copies of the suspension ordered by the Prefect and the summons to appear before the MEEA, if applicable.
3	Restoring the site to its original state	LSM Gold Corporation	SOME Benild	Immediate	Restore the flow of the blocked "Poni" river, fill in the excavations
4	Planting to compensate for damaged vegetation	LSM Gold Corporation	SOME Benild	July to September 2024	

CONCLUSION

The mission provided an opportunity to inspect the activities of LSM Gold Corporation, note the damage, take samples for analysis and propose solutions to resolve the crisis.

Appendix 1: Composition of the mission

N°	Full name	Structure	Contact
	BELEMSOGO Aristide	MEMC	71 50 79 79
	SAWADOGO Hyppolite	DCRP/MEMC	
	BAKO Adolphe	DGRE	70 03 84 04
	SAMA W. Pierre	DGRE	
	SAAOGO Hilaire	ECD/Poni	51 00 27 28
	ZON Hamidou	DCRP/MEMC	
	N'DO Joas Wenceslas	DGPE/DLAQE	76 52 68 01

Appendix 2: List of people met

N°	Full name	Structure	Contact
1	SENI D. Bienvenu	PDS/Gomblora	
2	GUENGUERE Lucien	Haut- comm/Poni	

Appendix 3: Images of the sites

SANY

PART IV

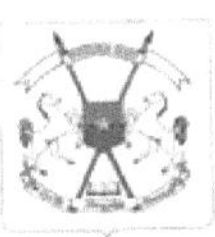

BURKINA FASO

Unité - Progrès - Justice

MINISTERE DE L'ENVIRONNEMENT,
DE L'EAU ET DE L'ASSAINISSEMENT

SECRETARIAT GENERAL

BURKINA FASO

Unity - Progress - Justice

MINISTRY OF THE ENVIRONMENT,
WATER AND SANITATION
GENERAL SECRETARIAT

MISSION REPORT

INVESTIGATION INTO THE DEATH OF
OF FISH IN THE KORSIMORO DAM
KORSIMORO DAM, CENTRAL NORTH REGION

March 2024

INTRODUCTION

In Burkina Faso, the preservation and sustainable use of water resources is one of the central pillars of development policy. Numerous projects and programmes have been implemented to improve access to this resource, through the construction of groundwater catchment works and surface water reservoirs. However, this initiative faces a twofold challenge: meeting the growing demand for water while coping with the impact of climate change.

In this context, project beneficiaries are increasingly integrating the use of agricultural inputs to increase their production and cope with climatic hazards. Although this practice is necessary to increase yields, it presents risks for the aquatic ecosystem, particularly due to the misuse of organic or mineral fertilisers and pesticides (Zongo, A. et al. 2018).

Surpluses of these products, carried by run-off water, accumulate in surface water reservoirs, leading to eutrophication that is detrimental to their ecological balance. Inputs of elements such as phosphorus, nitrogen and sulphur contribute significantly to this phenomenon. In addition, other risks are emerging, including pathogenic contamination due to flooding and increased pollutant concentrations during periods of drought (Zongo, A. et al. 2018).

The problem of polluted water resources has worsened in recent years in Burkina Faso, due to the use of chemicals, particularly cyanides, in mining and gold panning activities. This situation raises major concerns about the preservation of water quality and public health in the country.

Every year, pollution incidents are reported across the country in surface water courses and reservoirs. These pollution incidents have resulted in the death of animals and fish.

Recentlythe President of the Interim Special Delegation of the commune of Korsimoro, Centre-Nord region, sent a memo to the population of the said commune concerning cases of fish deaths at the Korsimoro dam on 12, 13 and 14 March 2024.

On the instructions of the Secretary General of the Ministry of the Environment, Water and Sanitation, a joint mission led by agents from the Directorate General for Water Resources (DGRE) and the Directorate General for the Preservation of the Environment (DGPE) travelled to the Centre-North Region, to Korsimoro to be precise, on 15 March 2024, to carry out technical investigations into the death of fish in the village dam.

The general objective of the mission was to determine the probable causes of this fish mortality in the Korsimoro dam.

The specific objectives of the mission were to :

- Meet the authorities at regional and/or local level and keep them informed
the objective of the mission ;
- Carry out direct on-site pollution surveys at the dam and
by interviewing local residents;
- Carrying out in-situ analyses and taking water samples for analysis.
in-depth laboratory analyses ;

The expected results were :

- The authorities at regional and/or local level have been informed of the objective.
of the mission ;
- On-site inspections of the dam were carried out and the local residents
were asked about the situation;
- In-situ analyses and taking water samples for analysis
were carried out in the laboratory;

1. OVERVIEW OF THE MISSION

The mission team first went to Korsimoro town hall, where it met the Chairman of the interim Special Delegation, Mr OUEDRAOGO Seydou, to discuss the purpose of the mission. Here is a summary of the information gathered during these discussions:

- Mr OUEDRAOGO Seydou has not yet visited the site himself.
- He described the incident as a case of poisoning.
- Witnesses at the site observed bottles of pesticides around the dam.
- Fish farming is traditionally practised in the bed of the dam by a group of fishermen, using floating structures or pens.
- The first signs of fish mortality were noted on 12 March 2024, but in insignificant numbers.
- On 14 March 2024, the number of dead fish increased significantly.
- The dead fish were not eaten, but were incinerated under the supervision of the local authorities and police.
- Samples of fish, mud and water were taken by various technical departments, including the Ministry of Fisheries Resources, the Regional Directorate of Water and Sanitation, and the National Agency for Environmental, Food, Labour and Health Product Safety (ANSSEAT, formerly LNSP);
- Market gardening is also practised around the dam.
- The incident appears to be a first for the dam, according to the information provided. The mission team **then** met with the Secretary General of the Korsimoro town hall, Mr Ange LANKOANDE, who was impatiently awaiting the arrival of the governor in order to visit the dam. A few minutes later, the representative of the governor of the Centre-North region, who was the High Commissioner of the Sanmatenga Province, arrived. After the courtesy greetings, the mission moved on to the dam site, accompanied by the Korsimoro town hall authorities and the High Commissioner.

Once on the site, we found several stakeholders **(Appendix 1)** who had come for the same cause. Discussions with some of them, such as the interim prefect of Korsimoro and the president of the traditional fishermen's cooperative, enabled us to gather the same information as that obtained from the Korsimoro town hall.

But what seems inconsistent in their explanations is that the fishermen claimed that the fish kill did not affect all the floating huts, but only a few that were in the same body of water close to each other and sharing the same water. What's more, the incident concerned large fish (those weighing more than 300g).

A report from the Sanmatenga Departmental Service for Agriculture, Animal Resources and Fisheries in Korsimoro gave further details:

- Fish mortality at the dam began on 12 March 2024 with a small number of fish and the highest mortality recorded was on 14 March 2024 ;
- Mortality concerned large fish aged 05 months;
- On 14 March 2024, eight (08) floating huts and four (04) enclosures were affected by mortality;
- The fish in the other floating huts and pens in the middle of the dam were still intact, and no mortality was observed;
- More than 2,752 dead Oreochromis niloticus were counted belonging to three (03) fish farmers.
- The dead fish were taken to another site for incineration, as recorded in the minutes.

In the field, the joint mission team was able to make the following observations:

- A high density of market gardening on the banks of the dam and around the reservoir, despite the measures prohibiting occupation;
- Gold panning in the area, especially towards the median-east side of the dam,
- Installation of swivel pumps on either side of the dam to make water available for market gardening;
- Granite crushing machines searching for gold all around the dam;
- The water in the dam is not very turbid;
- Presence of pens and floating huts on the water of the dam used as cages for the fish farm (**Appendix 2**) ;
- Movement of live fish in certain cages at the scene of the incident;
- The live fish are of the same species (*Oreochromis niloticus*) as the fish;
- A few pieces of dead fish were found on the ground on the left bank near the scene of the incident;
- Digging holes for gold on the median side of the right bank;
- No invasive plants in the dam;
- A considerable drop in the water level ;
- The reservoir began silt up (mounds of earth in the middle of the reservoir).

On the strength of these findings, some local residents are accusing the fishermen of using pesticide-contaminated bottles to collect water or feed fish. A 25kg bag filled with pesticide bottles was presented to us as evidence. Or the food supplied to the fish was of poor quality or even toxic. Finally, a case of criminality or score-settling through the use of poison or contaminated containers was also suspected.

Finally, after these routine observations, the team took water samples and carried out *in situ* measurements on both banks of the dam.

Three sampling points (**Appendix 2**) were identified and coded (P1: left bank: location 1 of the incident, P2: left bank: location 2 of the incident and P3: right bank on the median side: location of gold-panning digs). Their GPS coordinates were taken.

The water samples were taken in accordance with the Burkinabe standards "NBF 08-003-1 to 4: 2018" relating to water sampling, including surface water.

The water samples taken are kept cool in coolers. Some are then sent to the DGRE's Raw Water Analysis Laboratory (to analyse additional parameters such as major ions and trace metal elements), while others are sent to the Environmental Quality Analysis Laboratory (for BOD5 and COD analyses).

2. ANALYSIS METHODOLOGY

- 2.1. In situ analysis

The in situ measurements covered the following parameters: temperature, pH, conductivity, turbidity and dissolved oxygen.

- 2.2. Laboratory analysis

Laboratory analyses were carried out on :

- Major ions: nitrates (NO_3^-), nitrites (NO_2-), ortho-phosphates (HPO_4^{2-}), as well as sodium (Na) and potassium (K),
- Metalloids (metallic trace elements): copper (Cu), zinc (Zn), Manganese (Mn), Iron (Fe) ;
- The chemical indices (Biological Oxygen Demand (BOD_5) and Biochemical Oxygen Demand (BOD)) are calculated on the basis of the following parameters

Chemical Oxygen Demand (COD)).

- **2.3 Analysis equipment and methods used**

The equipment and methods used to analyse the parameters studied are listed in the table in Appendix 4.

3. RESULTS AND INTERPRETATION

3.1. Presentation of analysis results

Table 1 gives the results of in-tu and laboratory analyses of several physicochemical parameters.

Table 8: Results of in situ measurements and laboratory analysis of water samples

Samples		*Analysis results*																	
Cade samples	Geographical location	Temperature (°C)	pH	Conduct ;vity (pS/cm)	Turbidite (NTO)	Dissolved oxygen (mg/l)	Sodium Na (mg/l)	Potassium K (mg/l)	Copper Cu (mg/l)	Manganese Mn (mg/l)	Iron Fe (mg/l)	Zinc Zn (mg/l)	Cyanides GN' (mg/l)	Ammonium NH4+ (mg/l)	Nitrite (mg/l)	Drtho-Phosphate HPD43- (mg/L)	COD (mg/L) 02
PI (Incident location 1)	I2°49'33.9 "N DDI°D2'5D,9 "W	33,9	9,90	137,3	214,9	4,90	10,55	1,95	<0,05	0,25	0,91	<0,05	0,009	0,90	0,020	0,19	39
P2 (Lieo 2 of ('incident)	I2°49'32, W "N D DI°D 2'51,2 "W	33,2	7,41	135,9	130,9	9,49	9,39	3,23	<0,05	0,19	0,57	<0,05	0,009	0,74	0,049	0,23	50
P3 (Gold panning area)	I2°49'34, D "N DDI°D2'2 2.9 "W	33,9	9,05	135,7	134,9	7,24	3,72	5	<0,05	<0,05	0,91	<0,05	0,014	0,95	0,093	0,03	II
National Standards[1]		-	9à 10,3	IODO	-	>5	-	-	0,3	-	-	<1	-	<1	<0,03	-	-
Moroccan standards[5]		9<T<3D	5 à 9	<3000	-	>3	-	-	<40	<0,1		<1,3	<0,05	<1	<0,5	-	<30

[1] **Decree no. 2001-185/PRES/PM/MEE setting standards for the discharge of pollutants into the air, water and soil, Article 9: defining standards for the protection of fish water quality.**

3.2. Interpretation of analysis results

3.2.1. Temperature

The temperature of water affects its density and viscosity, the solubility of oxygen and the speed of chemical and biochemical reactions.

Temperatures varied between 33.2°C and 33.9°C, which is consistent with a normal temperature for an aquatic environment for *Oreochromis niloticus*.

This increase in temperature may be due to various factors, such as climate change, human activity or the period during which the fish are harvested. Nile tilapias (*Oreochromis niloticus*) prefer relatively warm water temperatures, generally between 25°C and 32°C. However, they can tolerate a wider range of temperatures, from around 10°C to 35°C,

[5] **Joint decree of the Ministry of Spatial Planning, Water and the Environment n°2028-03 (5 November 2003) setting the quality standards for fish water in the Maree.**

although their growth and reproduction are optimal within the aforementioned range **(Mohamed Saad El- Sherif and al.; 2009).**

3.2.2. pH

As far as pH is concerned, it varies between 6.90 and 8.05. A neutral pH is around 7, so the values are slightly acidic to slightly basic, but remain within an acceptable range for most aquatic ecosystems. In fact, tilapia can thrive in a wide pH range from 5 to 11. However, a pH range of between 7 and 8.5 is recommended. Between 6 and 9, there are no accidents by direct effect but a toxic action due to ammonia can be found.

3.2.3. Conductivity

Electrical conductivity is a measure of the ability of an aqueous solution to conduct an electric current, which depends on the concentration of ions in solution. It is sensitive to variations in dissolved matter, mainly mineral salts, and can be useful for characterising a body of water. Conductivity also increases with water temperature.

Conductivity range from 135.7 to 137.3 µS/cm. Conductivity measures the ability of water to conduct an electric current, which is influenced by the presence of dissolved ions. These values appear to be relatively stable and within an acceptable range.

3.2.4. Turbidity

The turbidity of the water depends on the nature of the terrain it crosses, the season, the rainfall, the water flow regime and the nature of the discharges. It can be assessed by the transparency of the water.

The values are quite high, ranging from 130.9 to 214.8 NTU, which suggests a certain presence of suspended particles in the samples. This may be explained by the fact that the water is stirred up every day by fish farmers travelling by pirogue to their cages.

3.2.5. Dissolved oxygen

Along with pH values, dissolved oxygen concentrations are one of the most important water quality parameters for aquatic life.

- It generally accounts for 35% of the volume of total gases dissolved in water.
- Air-water exchanges take place through contact increased by mixing on facies.

lotic. Photosynthesis increases dissolved oxygen levels. A level of less than 3 mg/l is abnormal.

- It is lethal for cyprinids when it falls below 4 mg/l and 6 mg/l respectively.
- g/l (75% saturation) for salmonids. The vital value for these fish must be greater than 6 mg/l.
- Levels above the natural oxygen saturation level indicate

eutrophication of the environment, resulting in intense photosynthetic activity.

Dissolved oxygen is essential for the vast majority of organisms living in water, as the solubility of O2 varies according to temperature, mineralisation and atmospheric pressure.

Dissolved oxygen levels range from 4.90 to 7.24 mg/l, indicating a more or less adequate quantity of oxygen for aquatic life.

3.2.6. Metallic elements (mg/l): Concentrations of metals such as copper, manganese, iron and zinc are all below detection limits (noted as <0.05 mg/l), except for potassium which varies between 3.72 and 10.55 mg/l. However, these values remain within acceptable limits according to national and Moroccan standards.

3.2.7. COD (chemical oxygen demand): Values range from 11 to 50 mg/L $_{O2}$. The values at the incident site are higher than the reference standard (< 30 mg/L). This could indicate organic pollution.

3.2.8. Other chemical parameters: Concentrations of cyanides, ammonium, nitrites and orthophosphates also comply with national and/or Moroccan standards.

In summary, the data show generally acceptable environmental conditions, although turbidity and COD are slightly elevated in some samples indicating traces of organic pollution that may have taken place. However, ongoing monitoring would be required to ensure that these conditions remain stable and do not exceed the limits recommended to preserve the health of aquatic ecosystems and human safety. Another case to consider is fish feed.

To do this, it would be useful to have the results of analyses carried out by other players who have also been involved, in particular the ministry responsible for fisheries resources and the ministry responsible for health through the Agence Nationale de Sécurité Sanitaire de l'Environnement, de l'Alimentation et du Travail et des produits de santé (ANSSEAT, formerly the Laboratoire National de Santé Publique), for more complete technical investigations.

The following recommendations (**Table II**) have been formulated in order to limit possible pollution of water resources throughout the country.

Table 9: Recommendations for quality management of water resources

N°	Recommendations	Deadline for implementation	Implementation structure
1	Implement a communication plan on the impact of human activities on water quality, and the interaction between the hydrographic network and the pollution of water bodies.	December 2024	Water agencies, DREA, CLE
2	-Prohibit the fish farming technique used by the cooperative. As the cages are immersed directly in the water, the food residues from the fish could cause organic pollution. Train fish farmers in fish farming techniques that would not present any risk to the quality of the dam's water.	Continue	Water Police, MARAH, CLE, Town Hall
3	Raise awareness among market gardeners of the need to comply with standards for occupying riverbanks and for producing and, above all, using pesticides and synthetic chemical fertilisers.	Continue	Water police, CLE, town halls
4	Bound the dam with dikes and create herbaceous strips around the water bodies, then plant non-woody forest plants around these strips to limit pollution of the water bodies.	Continue	Water agencies; CLE; Town hall
5	Ensure compliance with water regulations by up inspections	Continue	DREA (Service Water Police)
6	Strengthening water quality control and monitoring systems (water quality monitoring networks, strengthening the technical facilities of DGRE and DGPE laboratories).	Continue	MEEA, DGRE, DGPE
7	Given the recurrence of water pollution events, set up an Interministerial Technical Committee to investigate water pollution under the coordination of the MEEA.	December 2024	SG/MEEA
8	There is also a need revise the regulatory texts setting	December 2025	MEEA, MARAH;

	the standards for surface water/groundwater in general and fish water in particular, to make them national "NBF" standards for use in the following areas		ABNORM, Ministry of Health

CONCLUSION

The Korsimoro dam is located in the village of Koupela, a few kilometres east of Korsimoro, in the Centre-Nord region. Market gardening and gold panning are practised around the dam by local people, who use synthetic chemicals that are likely to pollute the water in the dam.

The aim of our mission was to determine the nature of any pollution of this reservoir by various pollutants in order to pinpoint the probable causes of the fish kill.

The mission team analysed the various physico-chemical parameters of the dam water *in situ* and in the DGRE and DGPE laboratories. The *in-situ* measurements and the results of the laboratory analyses indicate that the water samples analysed comply to a greater or lesser extent with the standards for the parameters studied. The data show generally acceptable environmental conditions, although turbidity and COD are slightly elevated in some samples, indicating traces of organic pollution. However, ongoing monitoring would be required to ensure that these conditions remain stable and do not exceed the limits recommended to preserve the health of aquatic ecosystems and human safety.

The fish farming technique used by the cooperative should be avoided. As the cages are immersed directly in the water, the food residues from the fish could cause organic pollution. These fish farmers need to be trained in fish farming techniques that do not present a risk to the quality of the water in the dam.

BIBLIOGRAPHICAL REFERENCES

1. Decree no. 2001-185/PRES/PM/MEE setting standards for pollutant discharges into the air, water and soil, and defining standards for the protection fish water quality.
2. Joint decree of the Ministry of Spatial Planning, Water and the Environment n°2028-03 (5 November 2003) setting quality standards for fish waters in Morocco ;
3. NBF 08-003-2 : 2018 : Water quality - Sampling - Part 2 : Storage and handling of water samples ;
4. NBF 08-003-4: 2018: Water Quality - Sampling - Part 4: Guidance for sampling the waters of natural and artificial lakes;
5. Mohamed Saad El- Sherif and Amal Mohamed Ibraim El-Feky, 2009, Performance of Nile tilapia (Oreochromis niloticus) fingerlings. Influence of different Water Temperatures, International Journal of Agriculture & Biology, 11:301-305).
6. Zongo, A., Ouédraogo, I., Traoré, B. (2018). Challenges to the preservation and sustainable use of water resources in Burkina Faso: The case water catchment and retention projects. Revue burkinabè de développement rural, 20 (2), 45-60.

APPENDIX 1: LIST OF AUTHORITIES MET IN THE LOCALITY

Surname and first name(s)	Function and/or Structure	Contacts
OUEDRAOGO Seydou, Naaba Sanem	Acting Chairman of the Special Delegation, Korsimoro Town Hall	70 25 91 21
LANKOANDE ANGE	Secretary General of Korsimoro Town Hall	Tel : 70 25 93 15
GAMSONRE Idrissa	High Commissioner for the Province of Sanmatenga	Tel : 70 81 05 43 / 74 77 90 57
LOMPO Madjoa	Acting Prefect of Korsimoro	Tel : 70 53 07 42
YAMEOGO Gael	Regional Director, DREA-Centre Nord	Tel : 76 50 86 68
PELEGSONRE Issiaka	President of the Fishermen/ Relwendé Cooperative	-
ZABSONRE Grégoire	Service departmental of the Korsimoro environment	75 51 63 83

Figure 1: Dry plan of the dam with mart fish debris

Figure 2: floating huts and encias used for fish farming in the dam's water body

Figures 3 and 4: niaraicheculture in the dam bed
Figure 5: Pesticide bottles grooved around the dam
Figure 6: Irons digging around the dam in search of gold

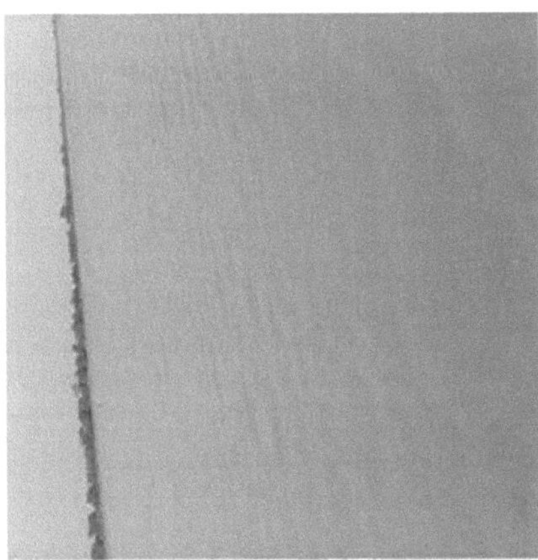

Figure 8: Profile view of the part of the dam not used by fishermen Figure 9: Water samples taken for laboratory analysis

Categories	Parameters	Description/interest
Classical physico-chemical parameters/ In-situ measurements	Temperature	Influences chemical equilibrium,
	pH	Measures acidity; distinction between acidic and basic water; linked the nature of the aquifer and environmental conditions
	Conductivity	Indicates the level mineralisation of the water
	Dissolved oxygen and saturation percentage	Determines the biological activity of the environment and its pollution status
	Turbidity	Provides information on water clarity
Organoleptic parameters	Colour	Tells you the colour of the water
	Odour	Tells you what the water smells like
Major elements, metallic trace elements	Potassium	Measures potassium content; indicator of mineralisation
	Sodium	Measures sodium content; indicator of leaching from aquifers containing silicates or sodium chloride
	Chlorides	Measures the content of these elements and provides information on the nature of the aquifer and water pollution
	Nitrates	
	Nitrites	
	Ortho phosphates	
	Sulphates	
	Free cyanides	Measures free cyanide content; provides information on contamination of industrial or commercial wastewater and mining effluent
	Total cyanides	Measures total cyanide content; provides information on contamination of industrial or commercial wastewater and mining effluent
	Copper	Measures copper content; Present in surface water or groundwater; its source is rock or industrial or mining effluents.
	Zinc	Measures Zinc content; Present in surface water or groundwater; its source is rock or industrial or mining effluents.
	Manganese	Measures manganese content; its source is rock or pollution caused by human activity
Organic pollution parameters	COD	Measures the oxygen content required to chemically oxidise organic compounds; Used to assess the quantity of biodegradable and non-biodegradable organic pollutants in surface water.
	BOD5	Measures the oxygen content required to biologically oxidise organic compounds; Used to assess the biodegradable fraction of the carbonaceous pollutant load in surface water within 5 days.
	Oxidability with KMnO4	Measures all oxidisable compounds (C , P N, etc ;)

Parameters analysed	Units	Methods	Standards	Equipment used	Detection limits
Temperature	°C	Electrochemical	NF EN 27888	Conductivity meter 3210	**0.0 à 60.0**
pH		Electrochemical	NF T 90-008	Hanna HI 98128	**-2.00 à 16.00**
Conducti vity	pS/cm	Electrochemical	NF EN 27888	Conductivity meter 3210	**0.0 à 3999**
Turbidity	NTU	Nephelometric	NF EN ISO	Turbidimeter 355 IR	**0.02**
Oxygen dis	mg/l	Electrochemical	-	Multimeter 3430 SET F	**0.02**
Chloride	mg/l	Chrom atography	HACH 8048	Metrohm IC ion	**0.02**
Nitrate	mg/l	Chrom atography	HACH Method	Metrohm IC ion	**0.3**
Phosphate	mg/l	Chrom atography	HACH Method	Metrohm IC ion	**0.02**
Sulphate	mg/l	Chrom atography	HACH 8051	Metrohm IC ion	**2**

Sodium	mg/l	Atomic absorption	FD T90-112	SAA Perkin Elmer Pin	**0.5**
Potassium	mg/l	Atomic absorption	FD T90-112	SAA Perkin Elmer Pin	**0.5**
Copper (Cu)	mg/l	Atomic absorption	FD T90-112	SAA Perkin Elmer Pin	**0.05**
Manganese	mg/L	Atomic absorption	FD T90-112	SAA Perkin Elmer Pin	**0.05**
Zinc (Zn)	mg/l	Atomic absorption	FD T90-112	SAA Perkin Elmer Pin	**0.05**
Cyanides	mg/l N-	Spectrophotometry (Pyrazalone-pyridine	-	DR 3900	**0,002**
Nitrites	mg/l	Spectrophotometry (Diazotation method)	-	DR 3900	**0,002**
Pesticides	mg/l	Gas chromatography method after liquid-liquid extraction.	NF EN ISO 6468/NF EN ISO 11369	GC Claros	**0.0100**

PART V

Unité - Progrès - Justice

MINISTERE DE L'ENVIRONNEMENT,
DE L'EAU ET DE L'ASSAINISSEMENT

SECRETARIAT GENERAL

BURKINA FASO

Unity - Progress - Justice

MINISTRY OF THE ENVIRONMENT,
WATER AND SANITATION

GENERAL SECRETARIAT

MISSION REPORT

INVESTIGATION INTO THE DEATH OF FISH
AT THE KALEINGA RESERVOIR IN THE GUIBA REGION
KALEINGA IN THE COMMUNE OF GUIBA SOUTH
CENTRAL REGION
CENTRE SUD

April 2024

INTRODUCTION

Managing water resources requires relevant information on both quantitative and qualitative aspects. Water quality, which used to receive very little attention, is now becoming a key issue in the integrated management of water resources. Pollutant discharges are increasing in number and diversity. These discharges, in context of inadequate sanitation, represent a real threat to public health and to the preservation of aquatic ecosystems. It is therefore necessary to monitor the quality of water resources and changes therein in order to gain a better understanding of current and future pollution, which may limit the quantity available and compromise various uses.

In Burkina Faso, quality monitoring of water resources is carried out at national level by the Ministry of Water, and more specifically by the Directorate General of Water Resources (DGRE) through the Water Quality Department, which has a laboratory for analysing surface and groundwater samples collected.

On 07 April 2024, a fish kill was recently observed at the Kalinga reservoir in the Guiba commune in the Centre-South region.

As a result, the Regional Director of Water and Sanitation for the Centre-South (DREA-CS) asked the DGRE to investigate the quality of the water in the Kalinga dam.

On the instructions of the Director General of Water Resources, a mission led by agents from the Directorate General of Water Resources (DGWR) visited the village of Kalinga on 15 April 2024 to carry out a technical investigation into the fish kill in the village dam.

The specific objectives of the mission were to :

- Meet the authorities at regional and/or local level and inform them of the mission's objective;
- Carrying out on-site inspections of the reservoir and interviewing local residents;
- Carrying out in situ analyses and taking water samples for in-depth laboratory analysis;

The expected results were :

- The authorities at regional and/or local level were informed of the purpose of the mission;
- On-site inspections of the reservoir were carried out and local residents were informed of the results.

asked about the situation;

- In-situ analyses and water sampling for in-depth analyses

in the laboratory.

1. OVERVIEW OF THE MISSION

The mission team **first** visited the DREA-CS in Manga, where it met the Regional Director of Water and Sanitation for the Centre-South, Mr KAFANDO Ouesseni and his colleagues, to discuss the purpose of the mission. Our discussions enabled us to gather the following information:

- The first fish kills were reported on 07 April 2024 and continued until 12 April, the date of their last visit to the reservoir;
- Fish mortality affected the entire surface of the reservoir;
- These are medium-sized or even small tilapia;
- The incident is a first for the dam," he said;
- There is no market gardening, gold panning or mining around the dam;
- Samples of fresh fish, mud and water were taken by

several technical departments, including: the ministry in charge of fisheries resources through the ZATE department and the Environmental Quality Analysis Laboratory.

We subsequently received a report from the DREA-CS water authority on the site of the incident. Here are the main points of the report:

- During our visit to the site, we observed an alarming situation where the basin of the reservoir was completely covered in dead fish, mainly tilapia of varying sizes.
- The water had a dark greenish tinge and an unpleasant, pungent smell.
- Crocodiles were observed paddling back and forth in the water, indicating a disturbance in the

ecosystem.
- The dam, originally intended for pastoral use, continued to be used intensively for watering animals locally and by nomads in transit with their animals between Burkina and the forest of Ghana, and vice versa.
- The use of wells located only 5 to 10 metres from the water level as a source of drinking water by local residents, despite the presence of dead fish floating in the dam.
- Animals, including sheep, donkeys, children and a woman, continued to drink the contaminated water.
- During our interviews, one child said that he was not ill and had never felt any pain, even when urinating.
- An older resident, who has lived in the area for more than 15 years, said that this was the first time such an event had occurred. He explained that farmers had come to exploit the dam, but that they had refused all agricultural practices in order avoid any pollution, leaving fishing as their only activity.
- The amount of water had fallen considerably, and although the dam is permanent and never runs dry, the dyke had not been maintained and was completely overgrown with trees and undergrowth.
- There are three large termite mounds on the dyke, one occupying almost half of the available space.
- Possible causes of the incident :
- It is possible that harmful chemical products
- weather conditions, particularly the recent hot and sunny weather, could contribute to a drop in oxygen in the water, given that it is shallow and stagnant. This could lead to fish suffocation and death;
- favourable environmental conditions, such as heat and stagnant water, could encourage the proliferation algae or bacteria, producing toxins that are harmful to fish;
- high heat and intense sunlight could contribute to stressful environmental conditions for fish, making them more vulnerable to disease or other mortality factors.

Lastly, mission team visited the Kalinga dam site, accompanied a team from DREA-CS.

Once on the site, we found the CVD and some local residents. Our discussions enabled us to gather the same information as that obtained from the Regional Water and Sanitation .

In the field, observations and interviews enabled us to gather the following information:
- The water is slightly greyish and not very turbid;
- The water from the dam gives off an odour;
- The DREA team has stated that there has been an improvement the quality of the water in the reservoir;
- Dead fish were found on both banks;
- These are medium-sized or even small tilapia;
- The incident was a first for the dam, they said, although one of them said (Appendix 1) that there had already been a similar case of fish deaths at the dam, but it only affected a portion of the dam and did not last long;
- Absence of invasive plants in the dam's water body ;
- No gold panning or mining sites in the vicinity of the reservoir;
- Absence of market garden crops around the dam;
- Small wells dug on either side of the reservoir serve as drinking water for local residents.
- Presence of birds and herds around the reservoir;
- Live crocodiles in the water ;
- No deaths of birds, mammals, lizards or amphibians, ...
- No incidents observed among the population;
- Presence of non-toxic plants (*Vitelaria paradoxa; Piliostigma sp;* etc.) around the incident site.

Finally, after these routine observations, the team took water samples and carried out in-situ measurements all along the stretch of water.

Four (04) sampling points (**Appendix 2**) were identified and coded (P1K (downstream); P2K: mid-point (in the middle of the water body); P3K (upstream) and P4K (well located upstream a few metres from the reservoir). Their GPS coordinates were taken.
Water samples were taken in accordance with Burkinabe standards NBF 08-003-1 to 4: 2018 on water sampling, including surface water.
The water samples taken are kept in coolers. They are then sent to the DGRE's Raw Water Analysis Laboratory for further analysis.

2. ANALYSIS METHODOLOGY

2.1. In situ analysis

The in situ measurements covered the following parameters: temperature, pH, conductivity, turbidity and dissolved oxygen.

2.2. Laboratory analysis

Laboratory analyses were carried out on :

- Major ions: nitrates (NO_3^-), nitrites (NO_2-); as well as sodium (Na) and potassium (K), ammonium (NH4+); cyanides (CN-);
- Metalloids (metallic trace elements and heavy metals): copper (Cu), zinc (Zn), manganese (Mn), iron (Fe); lead (Pb), arsenic (As), mercury (Hg).

2.3. Analysis equipment and methods used

The equipment and methods used to analyse the parameters studied are listed in the table in Appendix 4.

3. RESULTS AND INTERPRETATION

3.1. Presentation of analysis results

Table 1 gives the results of in-tu and laboratory analyses of several physicochemical parameters.

Table 10: Results of in situ measurements and laboratory analysis of water samples

Samples		Analysis results																	
Sample code	Geographical location	Temperato e (°C)	pH	Cond o ctivity (pS/c m)	Torbidit y (NTU)	Oxyge n e diesan e (mg/D	Sodio m Ne (mg/ D	Potassiu m mK (mg/D	Coppe r Co (mg/l)	Mang a nese Mn (mg/l)	Far Fa (mg/ l)	Zinc Zn (mg/l)	Cyanid e CM (mg/D	Ammoni o mNH4+ (mg/D	Nitrite (mg/D	Nitrat as (mg/ L)	Arseni c (mg/L) 02	Lead (mg/L)	Mercar e (ug/L
PI K (point ovo!)	12°49'33,9 "N 00Г02'50,Б" W	33,7	8,17	185,5	114,8	7,55	4,029	7,2	<0,05	0,895	1,07	0,089	<0,002	1,42	<0,002	<0,01	<0,001	0,017	<0,5
P2K (median point)	12°49'32,0 "N DD1ᴰD2'51,2 "W	33,2	8,24	183,8	139,4	7,95	18,18	5,29	<0,05	0,945	0,99	0,119	0,003	1,41	<0,002	<0,01	<0,001	0,015	<0,5
P3K (upstream point)	12°49'34,0 "N 001°02'22,9 "W	34,2	8,60	182,5	151,7	8,84	15,37	9,28	<0,05	0,990	1,28	0,043	<0,002	1,52	<0,002	<0,01	<0,001	0,019	<0,5
P4K (paite)	12°49'33,9 "N 00Г02'22,8TΛ 1	32	8,59	529	2Б	3,90	32,95	2,80	<0,05	6,53	0,Б8	0,109	0,005	2,32	<0,002	<0,01	<0,001	<0,001	<0,5
National Standards [1]		-	Бa 10. 3	1000	-	>7	-	-	0,3	-	-	<1	-	<1	<0,03	-	-	-	-
Moroccan standards[2]		8<T<3D	5 à 9	<3000	-	>3	-	-	<40	<0,1	-	<1,3	<0,05	<1	<0,5	-	<0,05	<0,02	<1

In red: concentrations exceeding standards

[3] **Décret n°2001-185/PRES/PM/MEE portant fixation de normes de rejets de polluants dans Γ air, l'eau et le sol, Artide 9 : définissant les normes pour la protecting the quality of fish waters**

[4] **Joint Order of the Ministry of Regional Planning, Water and the Environment no. 2028-03 (5 November 2003) setting quality standards for fish water in Morocco**

3.2. Interpretation of analysis results

For definitions of the various physico-chemical parameters, **appendix 3.**

3.2.1. In situ measurements

- **Temperature (°C)** :

The temperature of the water affects its density and viscosity, the solubility of oxygen and the rate of chemical and biochemical reactions. Temperatures appear to be relatively stable at the various sampling points, with minimal variations ranging from 32°C to 34.2°C.

- **pH** :

The pH is the cologarithm of the concentration of hydrogen ions in water: 0 to 7 is acidic and 7 to 14 is alkaline. The pH cannot be dissociated from temperature, dissolved oxygen and mineralisation. Fish generally tolerate a pH of between 6 and 9. The pH varies between 6.59 (P4K, well) and 8.6 (P3K, upstream point), indicating slight acidity in the well and higher alkalinity at the upstream point. The pH values are within the range recommended by national standards (6 to 10.3) but slightly above the upper limit of Moroccan standards (5 to 9) defining fish-farming water.

- **Conductivity ^S/cm)** :

Electrical conductivity (EC) is a measure of the ability of an aqueous solution to conduct an electric current, which depends on the concentration of ions in solution. It is sensitive to variations in dissolved matter, mainly mineral salts, and can be useful for characterising a body of water. Conductivity also increases with water temperature.

Electrical conductivity appears to be higher at the well (529 µS/cm) than at the other sampling points, where it is relatively stable at around 183-185.5 µS/cm. Conductivity is below the limit recommended by national standards (<1000 µS/cm) and well below Moroccan standards (<3000 µS/cm).

- **Turbidity (NTU)** :

The turbidity of the water depends on the nature of the terrain it crosses, the season, the rainfall, the water flow regime and the nature of the discharges. It can be assessed by the transparency of the water. Turbidity is relatively high at all points, with a slight increase from the downstream point to the upstream point, then a significant decrease at the well. Turbidity is high at all points, probably exceeding the recommended standards.

- **Dissolved oxygen (mg/l)** :

Along with pH values, dissolved oxygen concentrations are one of the most important water quality parameters for aquatic life.

- It generally accounts for 35% of the volume of total gases dissolved in water.
- Air-water exchanges take place through contact, increased by mixing on lotic facies. Photosynthesis at increases dissolved oxygen levels. A level of less than 3 mg/l is abnormal.
- It is lethal for cyprinids when it falls below 4 mg/l and 6 mg/l (75% saturation) for salmonids. The vital value for these fish must be greater than 6 mg/l.
- Levels above the natural oxygen saturation level indicate eutrophication of the environment, resulting in intense photosynthetic activity.

Dissolved oxygen is lower at the well (3.9 mg/l) than at the other sampling points, where it is relatively stable at around 7.55-8.84 mg/l. Dissolved oxygen complies with national standards (>7 mg/l) but is higher than the limit recommended by Moroccan standards (>3 mg/l) for water in the reservoir.

3.2.2. Major ions and metalloids

- **Sodium (Na)** :

Concentrations ranged from 4.029 mg/l to 32.95 mg/l. The highest concentration was observed at sampling point P4K, which appears to be a well.

- **Potassium (K)** :

Concentrations ranged from 2.8 mg/l to 9.28 mg/l. The highest concentration was observed at the upstream point (P3K).

- **Copper (Cu)** :

Concentrations were all below 0.05 mg/l in all samples, which is below the national standard of 0.3 mg/l and the Moroccan standard of <40 mg/l.

- **Manganese (Mn)** :

Concentrations ranged from 0.895 mg/l to 6.53 mg/l. The highest concentration was observed at the well (P4K). These concentrations exceed the limit value for the Moroccan standard, which is 0.1 mg/L, which could indicate potential contamination of the water in the reservoir. Indeed, the presence of manganese in the water may be a cause for concern for the health of fish and other aquatic organisms, as well as for humans who consume these fish. Manganese is a necessary trace element at low concentrations, but it can become toxic at high levels.

- **Iron (Fe)** :

Concentrations ranged from 0.68 mg/l to 1.28 mg/l. The highest concentration was observed at the upstream point (P3K).

- **Zinc (Zn)** :

Concentrations ranged from 0.043 mg/l to 0.116 mg/l. The highest concentration is observed at the mid-point (P2K). For zinc, the limit is <1 mg/l, also respected in all samples for national standards. The same applies to the Moroccan standard for zinc, with a limit of <1.3 mg/l, which was met in all samples.

In summary, analysis of the metal data shows significant variations in metal concentrations, with some exceeding national and Moroccan standards, especially manganese, which could indicate potential contamination of the water from the Kalinga dam. High concentrations of manganese can be toxic to fish, particularly over the long term. This can lead to gill damage, nervous system disturbances, reproductive problems and even mortality.

Manganese can accumulate in the tissues of fish over time, which can lead to an increase in the concentration of manganese in the aquatic food chain. Fish predators, including humans, can be exposed to high levels of manganese by consuming contaminated fish

High levels of manganese can also alter fish habitat by affecting water quality and reducing the availability of food resources.

It is crucial to regularly monitor manganese levels in fish waters and implement appropriate management measures to reduce manganese concentrations if necessary. This may involve water treatment practices or strategies to reduce sources of contamination.

The concentration of manganese in fish water is an important parameter to monitor because of its potential effects on the health of fish and aquatic ecosystems as a whole. Appropriate prevention and management measures are needed to minimise the risk of pollution and protect the health of fish populations and human consumers.

3.2.3. Nitrogen compounds

- **Cyanide (CN-)**: Concentrations are all below 0.005 mg/l, with most samples indicating a concentration below 0.002 mg/l. Point P4K (well) has the highest concentration at 0.005 mg/l. Cyanide concentrations in all samples are very low, well below national and Moroccan standards.
- **Ammonium (NH4+)**: Concentrations varied between 1.41 mg/l and 2.32 mg/l. Samples from point P4K (well) had the highest concentrations. Ammonium concentrations varied, with samples from point P4K (well) showing the highest concentrations. However, all concentrations remain above national and Moroccan standards.
- **Nitrite**: All samples show concentrations below 0.002 mg/l.
- **Nitrates**: All samples showed concentrations of less than 0.01 mg/l. Nitrite and nitrate concentrations are also very low in all samples, remaining below national and Moroccan standards.

In summary, cyanide, nitrite and nitrate concentrations in the water samples are negligible, while ammonium concentrations, although variable, are above national and Moroccan standards. It is important to monitor ammonium levels carefully, particularly in the samples from the well, where concentrations are highest.

3.2.4. Heavy metals :

- **Arsenic**: All concentrations were below 0.001 mg/L (or 1 µg/L). This means that arsenic concentrations are extremely low in all samples, well below national and Moroccan standards.

- **Lead**: Concentrations vary between <0.001 mg/L and 0.019 mg/L. Samples from point P3K (upstream point) have the highest concentration at 0.019 mg/L, while the other samples have concentrations below 0.02 mg/L, thus complying with national and Moroccan standards.
- **Mercury**: All concentrations were below 0.5 µg/L. This means that mercury concentrations are also very low in all samples, remaining below national and Moroccan standards.

Arsenic, lead and mercury are all toxic metals that can have harmful effects on human health at high concentrations.

In summary, the results of the heavy metal analyses show that the quality of the water in terms of the presence of arsenic, lead and mercury is generally good in the samples taken along the source, and they comply with national and Moroccan regulatory standards. However, ongoing monitoring is required to ensure that these levels remain low and do not pose a risk to human health and the environment.

4. RECOMMENDATIONS

The following recommendations (**Table II**) have been formulated in order to limit possible pollution of water resources throughout the country.

Table 11: Recommendations for quality management of water resources

N°	Recommendations	Deadline for implementation	Implementation structure
1	To raise awareness among the population living near the Kalinga reservoir about the use of harmful chemicals near water sources by prohibiting them from washing clothes directly in the reservoir to prevent any contamination.	Continue	DREA (Water Police Department)
2	Implement a communication plan on the impact of human activities on water quality, and the interaction between the hydrographic network and the pollution of water bodies.	December 2025	Water agencies, DREA, CLE
3	Ensure compliance with water regulations by stepping up controls	Continue	DREA (Service Water Police)
N°	Recommendations	Deadline for implementation	Implementation structure
4	Strengthening water quality control and monitoring system (water quality monitoring networks, strengthening the technical facilities of the DGRE laboratory).	Continue	MEEA, DGRE,
5	Given the recurrence of water pollution incidents, set up an Interministerial Technical Committee to investigate water pollution under the coordination of the MEEA.	December 2025	SG/MEEA
6	Revise the regulatory texts setting standards for surface water/groundwater in general and fish water in particular to make them national "NBF" standards for use	December 2025	MEEA, MARAH; ABNORM, Ministry of Health; Universities, Institut of Research, NGOs

CONCLUSION

The Kalinga dam is located to the south of the village, around 20 kilometres from the town of Manga in the Centre-Sud region. The local population, most of whom are Fulani, wash in the water using soaps or toiletries containing harmful substances that could contaminate the water and affect the health

of the fish;
The aim of our mission was to determine the nature of any pollution of this reservoir by various pollutants in order to pinpoint the probable causes of the fish kill.
The mission team analysed the various physico-chemical parameters of the dam water *in situ* and in the DGRE laboratory. It should be noted that the water sample was taken on 15 April 2024, more than a week after the incident.
The *in-situ* measurements and the results of the laboratory analyses indicate that the water samples analysed have characteristics that comply to a greater or lesser extent with the standards for the parameters studied. However, the concentration of manganese exceeds the limit value for the Moroccan standard, which is 0.1 mg/L for fish water. This indicates potential metalloid contamination of the water in the reservoir. High levels of manganese can alter fish habitat by affecting water quality and reducing the availability of food resources. The concentration of manganese in fish water is therefore an important parameter to monitor because of its potential effects on the health of fish and aquatic ecosystems as a whole. Appropriate prevention and management measures are needed to minimise the risks of pollution and protect the health of fish populations as well as that of human consumers.

BIBLIOGRAPHICAL REFERENCES

7. Decree no. 2001-185/PRES/PM/MEE setting standards for pollutant discharges into the air, water and soil, and defining standards for the protection of fish water quality.

8. Joint decree of the Ministry of Spatial Planning, Water and the Environment n°2028-03 (5 November 2003) setting quality standards for fish waters in Morocco ;

9. NBF 08-003-2 : 2018 : Water quality - Sampling - Part 2 : Storage and handling of water samples ;

10. NBF 08-003-4: 2018: Water Quality - Sampling - Part 4: Guidance for sampling the waters of natural and artificial lakes;

APPENDIX 1: A: Attendance list of the joint mission to the Kalinga dam

Order	Name	First name	Structures	Numbers
01	BAWAR	Barthélemy	DGRE	76072688
02	SEBGO W.	Amos	DGRE	73 06 20 69
03	ILBOUDO	Moumouni	DREA-CS	76 85 25 51
04	NOALI	Josué	DREA-CS	57 05 90 58

APPENDIX 1: B. List of authorities met at regional and local level.

Order	Name	First name	Function/Status	Numbers
01	KAFANDO	Ouesseni	Regional Director of DREA-CS	76641883
02	CONOMBO	Kuilga	CVD-Talinga	56 89 85 24
03	KAFANDO	Inoussa	Secretary (CVD- kalinga)	69 79 73 78
04	BATERO	Jean-David	Water and Forestry Officer	77 73 52 55/ 58 14 72 12
05	KABORE	Michel	Tracker	78 79 10 71
06	BAHADIO	Makioudou	Chief of the Peulhs	64 18 31 51
07	CONOMBO	Pascal	VDP-Kalinga	55 89 41 09
08	KABORE	Marcel	VDP-Kalinga	65 57 94 18
09	KAFANDO	Ousmane	VDP-Kalinga	68 51 61 63
10	DIALLO	Sadio	Kalinga national	68 45 70 54
11	BOUDA	Babayouré	Kalinga national	76 72 93 68

APPENDIX 2: ILLUSTRATIVE PHOTOS DGRE and DREA-CS, APRIL 2024

Figure 1: équipe de la mission sur la retenue d'eau figure 2: échantillons d'eau prélévé

Figure 3 : constat de poissons morts sur la rive du plan d'eau figure 4: troupeaux autour de la retenue d'eau

Figure 1: mission team on the reservoir figure 2: water samples taken
Figure 3: dead fish on the bank of the water body figure 4: herds around the reservoir

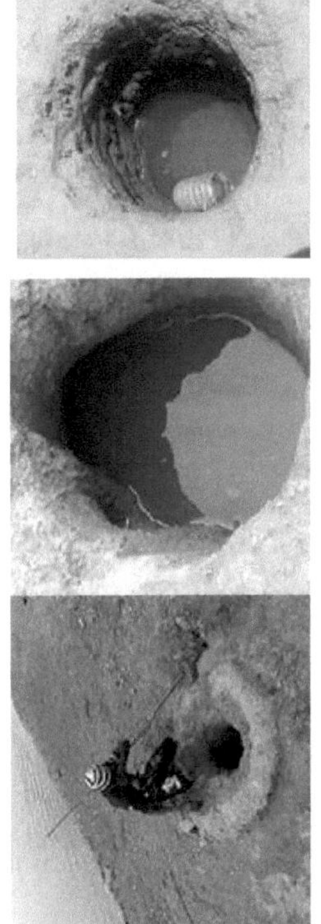

Figure 5 : puits à proximité de la retenue d'eau

Figure 6: Mesures in-situ

Figure 5: Well near the reservoir
Figure 6: In situ measurements

APPENDIX 3: TABLE I: DEFINITION OF SOME OF THE PARAMETERS STUDIED

Categories	Parameters	Description/interest
Conventional physico-chemical parameters/ In-situ measurements	Temperature	Influences chemical equilibrium.
	pH	Tacidity measurement; distinction between acidic and basic water; linked the nature of the aquifer and environmental conditions
	Conductivity	Indicates the level mineralisation of the water
	Dissonant oxygen and saturation percentage	Determines the biological activity of the environment and its pollution status
	Turbidite	Provides information on water clarity
Organoleptic parameters	Colour	Tells you the colour of the water
	Odour	Tells you what the water smells like
Major elements, metallic trace elements	Potassium	Measures potassium content; indicator of mineralisation
	Sodium	Measures sodium content; indicator of leaching from aquifers containing silicates or sodium chloride
	Chlorides	Measures the content of these elements and provides information on the nature of the aquifer and water pollution
	Nitrates	
	Nitrites	
	Ortho phosphates	
	Sulphates	
	Free cyanides	Measures free cyanide content; provides information on the contamination of industrial or commercial wastewater and mining effluent
	Total cyanides	Measures total cyanide content; provides information on contamination of industrial or commercial wastewater and mining effluent
	Copper	Measures copper content; Present in surface water or groundwater; its source is rock or industrial or mining effluents.
	Zinc	Measures Zinc content; Present in surface water or groundwater; its source is rock or industrial or mining effluents.
	Manganese	Measures the Manganese content; its source is the rock or pollution caused by human activities
Organic pollution parameters	COD	Measures the oxygen content required to chemically oxidise organic compounds; used to assess the quantity of biodegradable and non-biodegradable organic pollutants in surface water.
Categories	**Parameters**	**Description/interest**
	BOD5	Measures the oxygen content required to biologically oxidise organic compounds; Used to assess the biodegradable fraction of the carbonaceous pollutant load in surface water within 5 days.
	Oxidizability at KMηOι	Measures all oxidisable compounds (C, P, N, etc.)

APPENDIX 4: TABLE I: SOME OF THE ANALYSIS EQUIPMENT AND METHODS USED

Parameters analysed	Units	Methods	Standards	Equipment used	Detection limits
Temperature	°C	Electrochemical	NF EN 27888	3210 conductivity meter	**0.0 à 60.0**
pH		Electrochemical	NF T 90-008	Hanna HI 98128	**-2.00 à 16.00**
Conductivity	pS/cm	Electrochemical	NF EN 27888	3210 conductivity meter	**0.0 à 3999**
Turbidity	NTU	Nephelometry	NF EN ISO 7027	Wag-WT 3020 turbidimeter	**0.02**

Oxygen dissonance	mg/1	Electrochemical	-	Multimeter 3430 SET F	-
Chloride	mg/1	Chromatography	HACH 8048 method	Ion chromatograph	**0.02**
Nitrate	mg/1	Chromatography	HACH Method 8114	Ion chromatograph	**0.3**
Phosphate	mg/1	Chromatography	HACH Method 8008	Ion chromatograph	**0.02**
Sulphate	mg/1	Chromatography	HACH 8051 method	Ion chromatograph	**2**
Sodium	mg/1	Atomic absorption	FD T90-112	AAS	**0.5**
Potassium	mg/1	Atomic absorption	FD T90-112	AAS	**0.5**
Copper (Cu)	mg/1	Atomic absorption	FD T90-112	AAS	**0.05**
Manganese (Mη)	mg/L	Atomic absorption	FD T90-112	AAS	**0.05**
Parameters analysed	**Units**	**Methods**	**Standards**	**Equipment used**	**Detection limits**
Zinc (Zn)	mg/1	Atomic absorption	FD T90-112	AAS	**0.05**
Cyanides	mg/1 N-	Spectrometry (Pyrazalone-pyridine method)	-	DR3900	**0,002**
Nitrite	mg/1	Spectrometry (Diazotation method)	-	DR3900	**0,002**

Printed by Books on Demand GmbH, Norderstedt / Germany